T0276728

Current Research in Waste Disposal

Current Research in Waste Disposal

Edited by **Victor Bonn**

CLANRYE
INTERNATIONAL

New Jersey

Published by Clanrye International,
55 Van Reypen Street,
Jersey City, NJ 07306, USA
www.clanryeinternational.com

Current Research in Waste Disposal
Edited by Victor Bonn

International Standard Book Number: 978-1-63240-130-4 (Hardback)

Printed in the United States of America.

Contents

Preface   VII

Part 1   **Waste Disposal Site**   1

Chapter 1   **Geo-Environmental Site Investigation
for Municipal Solid Waste Disposal Sites**   3
Giulliana Mondelli, Heraldo Luiz Giacheti
and Vagner Roberto Elis

Chapter 2   **Application of Geophysical
Methods to Waste Disposal Studies**   31
Cristina Pomposiello, Cristina Dapeña,
Alicia Favetto and Pamela Boujon

Chapter 3   **Integrated Study
on the Distribution of Contamination
Flow Path at a Waste Disposal Site in Malaysia**   55
Kamarudin Samuding, Mohd Tadza Abdul Rahman,
Ismail Abustan, Lakam Mejus and Roslanzairi Mostapa

Part 2   **Nuclear Waste Disposal**   71

Chapter 4   **A New Generation of
Adsorbent Materials for Entrapping
and Immobilizing Highly Mobile Radionuclides**   73
Yifeng Wang, Huizhen Gao, Andy Miller and Phillip Pohl

Chapter 5   **Removal of Selected Benzothiazols with Ozone**   93
Jan Derco, Michal Melicher and Angelika Kassai

Chapter 6   **Modelling of Chemical
Alteration of Cement Materials in
Radioactive Waste Repository Environment**   119
Daisuke Sugiyama

Chapter 7   **Clarification of Adsorption
Reversibility on Granite that
Depends on Cesium Concentration**   145
Keita Okuyama and Kenji Noshita

Part 3   **Municipal Waste Disposal at Different Environments**   161

Chapter 8   **Assessment of Population Perception Impact
on Value-Added Solid Waste Disposal in Developing
Countries, a Case Study of Port Harcourt City, Nigeria**   163
Iheoma Mary Adekunle, Oke Oguns, Philip D. Shekwolo,
Augustine O. O. Igbuku and Olayinka O. Ogunkoya

Chapter 9   **Ballast Water and Sterilization of the Sea Water**   193
María del Carmen Mingorance Rodríguez

Part 4   **Emissions Related to Waste Disposal**   207

Chapter 10   **Evaluation of Replacing Natural Gas
Heat Plant with a Biomass Heat Plant – A Technical
Review of Greenhouse Gas Emission Trade-Offs**   209
James G. Droppo and Xiao-Ying Yu

Chapter 11   **Industrial Emission Treatment Technologies**   221
Manh Hoang and Anita J. Hill

Permissions

List of Contributors

Preface

This book has been a concerted effort by a group of academicians, researchers and scientists, who have contributed their research works for the realization of the book. This book has materialized in the wake of emerging advancements and innovations in this field. Therefore, the need of the hour was to compile all the required researches and disseminate the knowledge to a broad spectrum of people comprising of students, researchers and specialists of the field.

This book provides research-focused information regarding the methods and concepts of waste disposal. It presents research findings on various interesting issues in waste disposal. Topics such as geophysical processes in site studies, analysis of municipal waste disposal sites, study of contamination flow path at a waste disposal site, and case studies of disposal of municipal wastes in various environments and locations are discussed within the book. It also elaborates on issues regarding, emissions related to waste disposal and nuclear waste disposal.

At the end of the preface, I would like to thank the authors for their brilliant chapters and the publisher for guiding us all-through the making of the book till its final stage. Also, I would like to thank my family for providing the support and encouragement throughout my academic career and research projects.

Editor

Part 1

Waste Disposal Site

Geo-Environmental Site Investigation for Municipal Solid Waste Disposal Sites

Giulliana Mondelli, Heraldo Luiz Giacheti and Vagner Roberto Elis

Institute for Technological Research of São Paulo State,
São Paulo State University, University of São Paulo, São Paulo,
Brazil

1. Introduction

Deactivated dump sites and inadequate sanitary landfills can be a serious potential source of contamination. Due to the large number of these sites, contaminants can be generated and migrate into the ground. For this reason, site investigation programs which consider different types of soils and contaminants are needed. In many countries, sanitary landfills have been recently built according to the engineering standards with satisfactory procedures and following the requirements imposed by the local environmental agencies, including monitoring and proper operations. There is a great concern with the operation of sanitary landfills and with the future of deactivated dump sites located in small and medium-size cities, since many of them were inappropriately constructed and operated. Consequently, the main surficial streams and groundwater aquifers, subsoil and air become vulnerable to contamination and pollution. A continuous site investigation program, including in-situ and laboratory tests are necessary to identify the typical soil profile, hydrogeological characteristics and background chemical values. The site surroundings and the subsoil contamination plume can be characterized based on geotechnical, geochemical and mineralogical techniques.

The demand for the geo-environmental site investigation of contaminated and non-contaminated sites intended for future sanitary landfill planning has substantially increased in the last few years due to the lack of space in metropolitan areas.

Environmental agencies from several countries have proposed different site investigation methodologies in order to diagnose and confirm different contamination levels in sites with diverse physical characteristics in order to guide the remediation plan whenever it is necessary. The experience already achieved on site investigations has shown that the best methodology is site specific, and depends on subsoil and chemical contaminants; geotechnical, geological and hydrogeological aspects; evolution of the contamination plume and the possible risks it poses. Several field and laboratory investigation techniques (direct and indirect) have been proposed and used. Sometimes one technique is more suitable than another depending on the physical and natural characteristics of the site. Countries with recent concerns over the environment tend to adapt the experience gained from the more developed countries with the reality of their own environmental laws, economy, industrialization, size, etc.

Having outlined this scenario, this chapter aims to present and discuss the different tests and steps of a geo-environmental site investigation program proposed for municipal solid waste disposal sites.

2. Geo-environmental site investigation steps

The various steps that can be included during a geo-environmental site investigation are discussed in this item and can be systematized in the chart shown in Figure 1, which was prepared considering experience gained on several geo environmental site characterization campaigns at municipal solid waste disposal sites in Brazil. The major focus of these studies was to assess contamination of medium size municipal solid waste disposal sites, installed over typical Brazilian tropical soils (residual soils and sandstones).

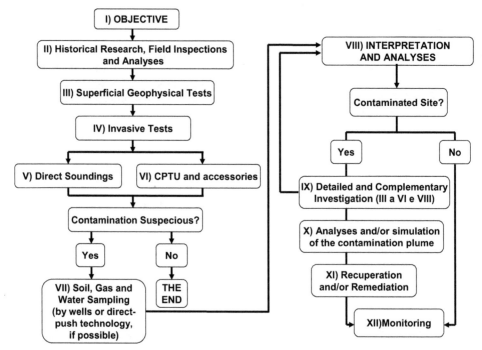

Fig. 1. Proposed geo-environmental site investigation steps for municipal solid waste disposal sites (Mondelli, 2008).

It is important to emphasize that the very first step in any site investigation program is to define the problem. For example, if the objective of the site investigation is just to characterize the local subsoil before the installation of a sanitary landfill, it will probably be necessary to carried out the site investigation until the fifth or sixth step presented in the Figure 1. Depending of the enterprise size or of the natural environmental and anthropological site conditions, the contamination will be assessed in a later step. Therefore, before starting any site investigation test it is fundamental to clearly define the objective of the site investigation program and determine the possible sources and/or targets of contamination. If the sources and/or targets do not exist, there is no sense in continuing the site investigation after the seventh step (Figure 1).

3. Preliminary investigation

Environmental agencies usually recommend gathering historical information and carrying out a site inspection during the preliminary investigation. Compilation of physical and historical data and all the existing information about the site need to be collected before starting any site investigation. This phase (second step in Figure 1) has proved to be very important for the future steps, based on the interpretation of the previous and current topography, hydrology, aerial photographs and historical documents. A good job in this phase can avoid unnecessary tests in the next steps of the site investigation, reducing costs without compromising the quality of the final results.

Before installing a new sanitary landfill, closing an old dump or in a regular landfill, or just to investigate the extension of a contamination plume caused by leachate from an irregular municipal solid waste disposal site, it is necessary to contact the responsible company and environmental agencies involved, visit the site and request the following data:

1. *Location*: geographical coordinates, original and current topography, distance from the urban zone, access, surroundings, rivers and airport proximity. Depending on the local regulations, some of these aspects may constrain the construction or extension of new sanitary landfills and may also help the site investigation plan and the final design.
2. *Municipal waste management*: information such as amounts of municipal, hospital, demolition/construction and sweeping/pruning waste generated per person and the locations and manner in which they are disposed of, also needs to be checked. Furthermore, some questions need to be addressed, like: "How long the current landfill will continue to operate? Has it obtained operating and/or commissioning licenses? Is there a plan for hazardous waste management? What percentage of the waste is recycled by the municipality? Are there any environmental education programs for the local community?"
3. *Aerial photography*: a very complete study of the site to be investigated, its uses and occupational activities and physical surrounds can be done by using aerial photography taken over a period of time up to present day photos.
4. *Old reports*: reports of past site investigations and geological surveys need to be available for consultation and search of the physical media data to be also used or supplemented during the new site investigation phase.
5. *Weather Aspects*: climate data, such as mean annual temperatures, rainfall, radiation, etc for the city or that region where the site investigation will be conducted, should be collected for analysis and calculation of the water balance. The water balance can be carried out using the Thornthwaite and Mather (1955) methodology.

After the analyses and interpretation of all integrated existing data from the site to be investigated, a *Preliminary Investigation Report* can be submitted. Therefore, the site investigation plan will be planned and conducted based on all information collected on site, facilitating the achievement of the final objective of the site investigation, in order to reduce time and costs. Sometimes, this next phase is called *Confirmatory Investigation*, when suspicion of contamination occurs. More recently, this phase can integrate the use of different field and laboratory tests. The most well known and widely used tests are surficial geophysics (non-invasive and indirect techniques), followed by piezocone tests (CPTU) (invasive and indirect/direct techniques) with special sensors and samplers for environmental purposes, and groundwater level monitoring wells. Accordingly with Figure 1, some of these techniques, which are more frequently used for geo-environmental site investigation purposes, will be briefly discussed.

4. Geophysical survey

Geophysical methods, particularly electrical techniques, can be used to study those different environmental characteristics which are important during the site characterization for waste disposal and the monitoring of migratory contamination plumes. The major advantage of these methods is that the measurements of soil and contaminant properties are indirect, but it can also be a disadvantage too. It allows a quick investigation of a large area avoiding direct contact with contaminants.

Direct techniques based on previous tests, such as simple reconnaissance boring with Standard Penetration Test (SPT) and monitoring through groundwater wells and piezometers, can be used and are considered a traditional practice in developing countries. In this context, much effort has been concentrated on measuring electrical resistivity, a property that is highly dependent on lithology and contamination, which are essential factors to be detected during a geo-environmental site investigation. The major interest in measuring electrical resistivity in a geo-environmental site investigation of contaminated sites is due to the fact that it is an indirect measurement that can be achieved by means of subsurface and/or surface tests, which is also sensitive to temporal variations of the physical environment. This also allows it to be used as a monitoring technique and not only as a method for the investigation and initial reconnaissance of the subsoil.

The following geological, geotechnical, hydrogeological and environmental characteristics can be assessed using geophysical methods: (a) rock depth; (b) discontinuities; (c) changes of soil texture; (d) groundwater level; (e) groundwater flow; (f) presence and three-dimensional distribution of waste; (g) contaminated soil; (h) contaminated groundwater and plume shape. The first five characteristics are essential for the assessment of any waste disposal sites. All these submitted characteristics are recommended in the geo-environmental monitoring of a waste disposal facility. The geophysical methods which present good results for these applications are the Resistivity and the Low Frequency Electromagnetics (Ground Conductivity Meters).

4.1 Resistivity

The Resistivity Method applies an artificial electrical current I, introducing two electrodes into the ground (A and B), with the objective of measuring the differential potential ΔV generated by two other electrodes (M and N) in the electrical current flow extremities (Figure 2). This arrangement allows the calculation of apparent resistivity ρ_a in subsurface, using the following equation:

$$\rho_a = K.\frac{\Delta V}{I} \text{ (ohm.m)} \tag{1}$$

where K is a geometrical factor that depends on the electrodes position in the ground and can be calculated for any electrode arrangement using the following general expression:

$$K = \frac{2\pi}{(1/AM)-(1/AN)-(1/BM)+(1/BN)} \text{ (m)} \tag{2}$$

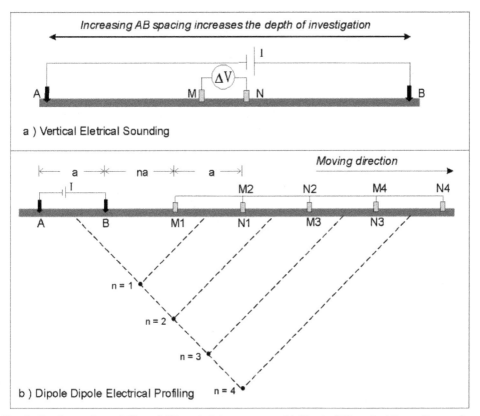

Fig. 2. Configuration of 1D and 2D resistivity surveys: (a) Vertical Electrical Sounding and (b) Dipole Dipole Electrical Profiling.

The soil and rock resistivity values are affected mainly by four factors: mineralogy composition, porosity, moisture, amount and nature of dissolved salts. The most important factors are the porous water content and the salinity content in the porous media. The increase of these two factors tends to decrease the resistivity values. These characteristics allow a good applicability of this geophysical method of application in environmental and hydrogeological studies, where inorganic contaminants occur. In the specific case of municipal solid waste disposal sites, the formation of leachate with elevated concentration of ions, characterizing the polluted areas with very low resistivity values.

The equipment used for the resistivity method consists basically of a controlled electrical current source emission and differential potential measurements. The source potential can range from some kilowatts up to hundreds of watts. This equipment can work using direct or alternating current supply with low frequency, preferably lower than 60 Hz (Telford et al., 1990).

The Resistivity Method encompasses various techniques for the application of field tests, which basically consist of vertical electrical sounding (1D resistivity survey) and electrical profiling (2D resistivity imaging) and includes a wide variety of possible electrode configurations (Schlumberger, Wenner, Dipole Dipole, Pole Dipole, Lee and other), rendering the method highly versatile. Currently, 3D tests are possible with the use of multi-electrode equipment or a series of 2D profiles grouped and equidistant (Loke & Barker, 1996; Dahlin et al., 2002). Figure 2 presents electrical soundings and dipole dipole electrical profiling field arrangements. As the subsoil is not homogeneous, for any measurement point an apparent resistivity value is obtained. Appropriate software is used for data acquisition and interpretation, for obtaining the electrical stratigraphy of the subsoil (1D survey) or the resistivity cross-section (2D survey). There are several possibilities for carrying out these tests allowing the use of different techniques and arrangements, depending on the site investigation objectives. Usually the most important objective of a geo-environmental characterization is to delineate contamination plumes.

The vertical electrical soundings can give important information on site characterization of waste disposal sites. The distribution of different geomaterials on the subsurface profile (several soil layers, waste and rock substrate top) and the depth of the saturated zone can be estimated using vertical electrical soundings. Figure 3 presents the results and the interpretation of a vertical electrical sounding carried out to expand the knowledge of the stratigraphy profile and for defining the position of the saturated zone. The combined results obtained from several vertical electrical soundings allow the construction of a contour map of the groundwater level showing the direction of the groundwater flow.

Waste disposal sites can be mapped using several vertical electrical soundings (VES) points or using electrical profiling (continuous way). The VES are more appropriate to vertical subsoil profile definition (waste thickness and landfill bottom). The electrical profiling gives a better delineation of the waste disposal dimensions. 2D inversion surveys of electrical profiling show both lateral and vertical variations of the materials in the cross-section. Figure 4 presents two municipal solid waste trenches excavated on sandy geological substrate delineated by electrical Resistivity Method results (resistivity values lower than 20 ohm.m characterize the places filled with municipal solid waste).

Figure 5 shows the results of a survey performed in a contaminated site downstream from a landfill in Brazil. This site was investigated using a 3-D resistivity imaging technique. The purpose of this investigation was to detect and delineate the contamination plume produced by the waste and to acquire detailed information about the affected area. The data set consisted of a series of parallel electrical profile data acquired with the dipole-dipole arrangement, and was inverted as a complete 3-D survey. The resistivity model identifies the disposed waste and the contaminant, marked by the isosurface values lower than 20 ohm.m. The results indicate the presence of a contamination plume and its preferred path. Monitoring wells were installed in the affected area and their chemical analysis confirmed the influence of contaminants.

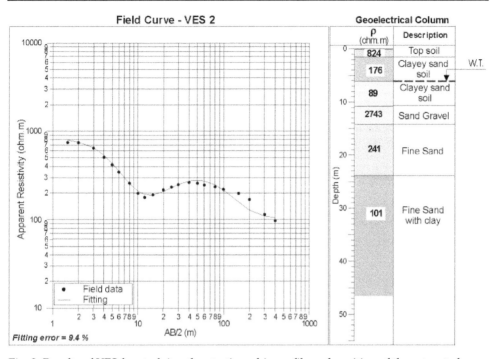

Fig. 3. Results of VES for studying the stratigraphic profile and position of the saturated zone (W.T.).

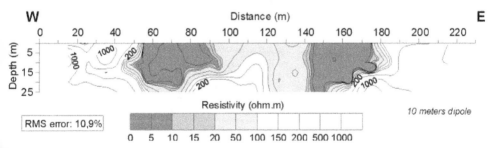

Fig. 4. Resistivity cross-section of waste disposal area. Lower resistivity zones characterize two trenches filled with municipal solid waste.

4.2 Low Frequency Electromagnetics (Ground Conductivity Meters)

The Electromagnetic Methods (EM) involves low frequency electromagnetic field propagation. When an alternating current (AC) is established in a wire placed on the ground surface, electrical currents flow in subsurface conductors. This process is known as electromagnetic induction and constitutes the basis of the operation of low frequency electromagnetic methods.

This method is very fast and it uses simple equipment easily operated. It explains the extensive application of this method for geo-environmental studies. The equipment used can be generally called ground conductivity meters. It consists of two coils (transmitter and receiver). The transmitter coil emits a primary magnetic field H_p, which induces electrical currents on the subsurface, generating a secondary magnetic field H_s (Figure 6). The combination of these two fields is measured by the receiver coil. Under certain conditions it is known that there is a linear relationship between the modules of the two fields (McNeil, 1980), technically defined as "low induction number operation". Accordingly, the apparent conductivity can be calculated by σ_a:

$$\sigma_a = \frac{4}{\omega\mu_0 s^2}\left(\frac{H_s}{H_p}\right) \quad (mS/m) \tag{3}$$

where ω = angular frequency, $2\pi f$; μ_0 = vacuum magnetic permeability; s = intercoil spacing.

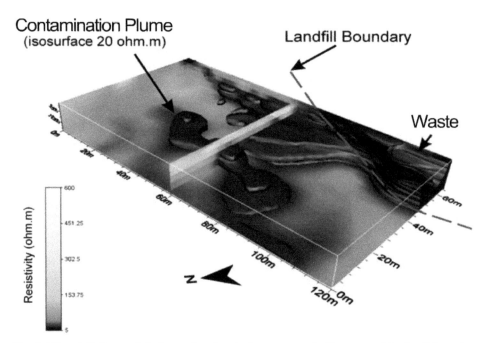

Fig. 5. 3D resistivity model of a contaminant plume generated by a municipal solid waste disposal site. The isosurface of 20 ohm.m delineates the contaminant plume, which is moving in a northwest direction.

The equipment is built to allow a direct reading of the apparent conductivity. The most commonly used equipment at present time was designed to explore pre-defined depths, between 7.5 to 60.0 meters, depending on the coils orientation (vertical loop and horizontal loop modes), the operating frequency and the intercoil spacing. The field tests are usually conducted in profiles, which, due to convenience of operation and transportation of the equipment, are carried out very quickly. The apparent conductivity data can be plotted on several profiles and depending on the distance, a set of profiles can form a site map. The interpretation of these data is basically qualitative, but nowadays there is inversion software designed for quantitative interpretation of conductivity profiling data (Monteiro Santos, 2004).

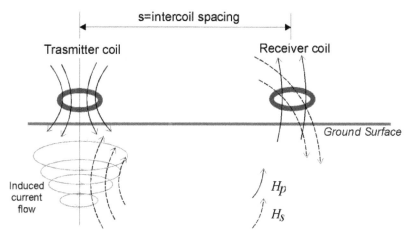

Fig. 6. Operating principle of the Ground Conductivity Meter.

The method is a cost effective and low time-consuming tool for investigating contamination generated by waste disposal sites, and allows delineation of the area affected by contaminants. Figure 7 shows the mapped out area of the contamination plume delineation, which has been created by a municipal solid waste disposal site. The MSW boundaries are marked in a broken white line (black points are measurement stations). The northwestern part of the landfill presents low electrical conductivity, characterizing a granitic base. The site is occupied by waste deposited onto tertiary stream sediments, and the groundwater level appears at around 5 meters depth. The contours of the waste disposed at the site can be identified based on the values of apparent electrical conductivity higher than 50 mS/m. The contamination plume can also be observed in this figure, which flows to the West, marked by high values of apparent conductivity outside of the waste filled area.

5. Invasive tests

In some particular sites it is difficult to identify the subsoil properties contrasts based only on geophysical data or index properties, especially in a very heterogeneous municipal solid waste (MSW) disposal site on unsaturated tropical soils. The proper site investigation in these sites relies on invasive tests, including soil, groundwater and direct gas sampling.

The modern approach for geo-environmental site characterization using invasive tests relies on the piezocone technology. The piezocone and it accessories are used as a screening tool for stratigraphic logging of geotechnical and chemical measurements. This approach often allows identifying potentially critical zones which may require more specific tests to measure or monitor contaminants based on specific sampling and/or monitoring.

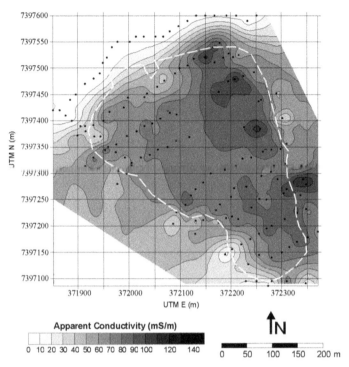

Fig. 7. Apparent conductivity map of a waste disposal site. Theoretical depth of exploration is 7.5 meters. The broken white line marks the limits of the deposit. Note the good match of the boundaries of the deposit with the high apparent conductivity values (> 50 mS/m).

5.1 Piezocone test and accessories

5.1.1 The piezocone

Piezocone penetration test (CPTU) is a standard site investigation tool currently used for geotechnical site characterization (Lunne et al, 1997). According to Shinn II & Bratton (1995), the heart of a modern site characterization approach uses the piezocone penetration test. The CPTU test uses a standard instrument probe (ASTM D 5778-07) with a 60° apex and typically 35.68 mm diameter (10 cm² area) fitted on the end of a series of rods (Figure 8). The probe is pushed into the ground at an approximate constant rate of 2 cm/sec by a hydraulic pushing source, such as a standard drill rig or cone pushing vehicle. As the cone is advanced, the forces measured by the tip and friction sleeve will vary with the material properties of the soil being penetrated. Tip resistance (q_c), friction sleeve stress (f_s) and dynamic pore-pressure (U) response are measured by calibrated electrical instruments. All channels are continuously monitored and typically digitized at 10, 20, 25 or 50 mm intervals.

5.1.2 Piezocone data and interpretation

Piezocone test data is gathered by a computer which allows the user to carry out straightforward post-investigative analysis. The three parameters (q_c, R_f and U) in various combinations, such as friction ratio ($R_f = (f_s/q_c)100\%$), are used to delineate site stratigraphy by using soil classification charts, as proposed by Robertson et al. (1986) and presented in Figure 9. Empirical and semi-theoretical correlations are available in relevant literature in order to estimate mechanical properties of the soil.

Measurement of equilibrium pore pressure at full pore pressure dissipation allows quantifying the vertical hydraulic gradients by using a single sounding and groundwater flow regime, whenever multiple soundings are available. The pore pressure dissipation test also allows estimation of hydraulic conductivity (Campanella, 2008).

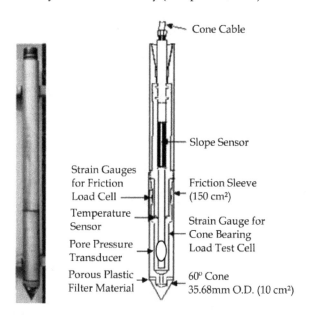

Fig. 8. Schematic drawing and photo of a piezocone probe (modified from Davies & Campanella, 1995).

5.1.3 Accessories

As pointed out by US EPA (1989), a geo-environmental site characterization requires information on the chemical distribution and sources (s)/receptor(s) for potential or existing contaminants. The piezocone technology for geo-environmental application included specific sensors for temperature, resistivity, pH, laser-induced fluorescence, among others. This technology also includes samplers to be used together with the piezocone for sampling soils, water and gas (Lunne et al, 1997). Robertson & Cabal (2008) present in details the main piezocone accessories available for geo-environmental site characterization.

The use of the piezocone test and accessories for geo-environmental applications essentially creates no cuttings, produces little disturbance, and reduces contact between field personnel and the contaminants, as the penetrometer push rods can be decontaminated during retrieval (Robertson, 1998). The major limitation of the piezocone test is the impossibility of carrying out the test in gravel.

5.1.4 Resistivity piezocone

Resistivity is one of the piezocone accessories which is called the resistivity piezocone (RCPTU) test. It measures the electrical resistance of a current flow in the ground. This additional information is extremely useful due to the significant effects that dissolved and free product constituents have on bulk soil resistivity (Campanella, 2008).

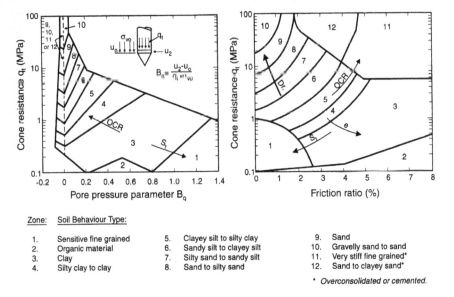

Zone:	Soil Behaviour Type:				
1.	Sensitive fine grained	5.	Clayey silt to silty clay	9.	Sand
2.	Organic material	6.	Sandy silt to clayey silt	10.	Gravelly sand to sand
3.	Clay	7.	Silty sand to sandy silt	11.	Very stiff fine grained*
4.	Silty clay to clay	8.	Sand to silty sand	12.	Sand to clayey sand*

* *Overconsolidated or cemented.*

Fig. 9. Piezocone soil behavior type charts proposed by Robertson et al (1986).

The preparations involved in the RCPTU test are similar to those of any other CPT test (Robertson & Campanella, 1988). The only additional procedure is the connection of a signal generator to the data acquisition system for controlling the current level and frequency of the electrical resistance measurements. Weemees (1990) discusses the importance of working with alternating current and sufficiently high frequencies to prevent polarization of the external electrodes. A frequency of 1000 Hz is usually used.

Figure 10 presents a Wenner-type resistivity piezocone (RCPTU) with an array of four electrodes. Resistivity measurements are taken with the inner electrodes and the current is applied through outer electrodes. These measurements are digitally recorded at 25 mm intervals, providing essentially continuous in-situ data sampling in addition to all the other standard CPTU measurements.

Measurements of bulk resistivity trends indicate whether some dissolved or free product exists below or above the background values. The background values are usually established from RCPTU tests carried out on-site. According to Campanella (2008), the areas where readings are very different (anomalies) from the background values are then further evaluated with appropriate groundwater sampling at discrete depths for detailed chemical analysis. Considerable practical value is gained from the fact that the measured resistivity in saturated soil is almost totally governed by the pore fluid chemistry.

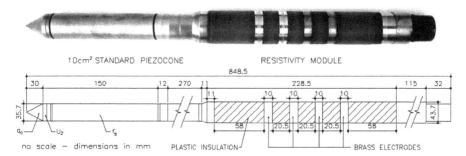

Fig. 10. Schematic representation and photo of a resistivity piezocone probe with a four-electrode arrangement (Mondelli et al., 2007).

5.2 Samplers

A large number of samplers for gas, water and soil sampling have been developed for pushing into the ground using the same equipment that is used for the piezocone test. These samplers are designed for either one-time samples or as monitoring wells. Robertson & Cabal (2008) describe in details some samplers available for geo-environmental site investigation and they will be briefly presented in the follow items.

5.2.1 Soil sampling

A wide variety of push-in discrete depth soil samplers are available and most of them are based on designs similar to the Gouda soil samplers. Robertson & Cabal (2008) describe some soil samplers including the Gouda type sampler. It is pushed into the ground at the desired depth in a closed position and it has an inner cone tip that is retracted to the locked position leaving a hollow sampler with small diameter (25 mm) stainless steel or brass sample tubes. The hollow sampler is then pushed to collect a sample. The filled sampler and push rods are then retrieved to the ground level. Figure 11 shows a schematic drawing of a typical CPT based soil sampler (Gouda type), as described in detail by Robertson & Cabal (2008).

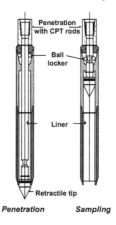

Fig. 11. Schematic drawing of a typical (Gouda type) CPT based soil sampler.

5.2.2 Water sampling

Robertson & Cabal (2008) affirm that the most common direct-push, discrete depth, in-situ water sampler is the Hydropunch and to a lesser extent, the BAT, Simulprobe and Waterloo Profiler. The Hydropunch sampler and its variations is a simple sampling tool that is pushed down to the desired depth and the push rods withdrawn to expose the filter screen and is described in detail by Robertson & Cabal (2008). A modification of the commercially available BAT groundwater sampler is recommended by Campanella (2008) for obtaining in-situ pore fluid samples and it is presented in Figure 12. This author discribes the original BAT system, which consists of a sampling tip that is accessed through sterile evacuated glass sample tubes and a double-ended hypodermic needle set-up pushed through septum seals. The tube sampler is lowered either by cable or electrical wire depending upon whether a pore fluid sample is taken with or without a pressure test being carried out. Acording to Campanella (2008), the BAT probe is also able to take pore gas samples for collecting volatile contaminants.

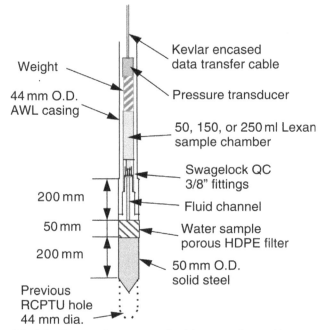

Fig. 12. UBC modified BAT groundwater sampler (Campanella, 2008).

5.2.3 Gas sampling

Gas samples can be collected in a similar way to the one previously described for groundwater samples. The most common gas sampler that uses the direct-push, discrete depth, in-situ method is the hydropunch type sampler. Robertson & Cabal (2008) also describes this sampler in detail, which is pushed to the required depth, thus the filter element is exposed and a vacuum is applied to draw a vapor sample to the surface. Special disposable plastic tubing is used to draw the sample to the surface.

5.3 Example of using RCPTU test and samplers in a MSW disposal site

An example of using the piezocone technology to assess contamination of a MSW disposal site from Brazil, by using different techniques is going to be briefly presented based on Mondelli's (2008) research. The site can be considered to be a controlled dump site because it is a planned landfill that incorporates some of the features of a sanitary landfill.

According to Mondelli et al. (2007), the site's geology has sandstone from the Adamantina and Marilia Formations, covered by alluvial sandy soils or colluvial clayey sands. Residual soils from sandstone are found underneath these layers. The hydraulic conductivity of the soil was found to vary from 10^{-7} to 10^{-6} m/s. The depth of the groundwater level is about 5 meters below the base of the landfill. In order to protect the subsurface and the shallow groundwater table from leakages from the landfill, four 20-cm-thick layers were compacted at the bottom of the landfill using local soil. The upper surface of these layers was coated twice with diluted asphalt emulsion to seal and protect the bottom of the landfill. This procedure was used because it was inexpensive and acceptable for controlled dumps in Brazil when this landfill was established.

This site was first investigated using a 3-D resistivity imaging technique, as described by Ustra et al. (2011). Several piezocone and resistivity piezocone tests were carried out on this site. In the first campaign, a four standard piezocone was carried out, measuring cone tip resistance (q_c), sleeve friction (f_s) and pore pressure (u_2). A resistivity piezocone was used in the other campaigns (16 tests). Pore pressure was recorded using a slot-filter filled with automotive grease in all the tests, as suggested by Larsson (1995).

A multifunction reaction system equipped with a hydraulic device with a capacity of 200 kN was used to carry out the tests. Details of all the different tests carried out in this site are presented by Mondelli et al. (2007). Two piezocone tests were conducted at the highest parts of the landfill, to obtain a reference profile of the uncontaminated soil. However, resistance to cone penetration was greater than the capacity of the penetration system needed for reaching the groundwater level in this region, where residual soils and sandstone are shallower. Most of the RCPTU tests were performed downstream of the landfill, where sedimentary soils or more homogeneous tropical soils occur.

All the RCPTU tests that reached groundwater presented an abrupt reduction in resistivity. This information was useful to help identifying groundwater level (GWL). Two of the RCPTU test results are presented in Figure 13 and will be briefly discussed. The resistivity profiles shown in this figure for the saturated zone, indicate that the resistivity measurements are affected by soil texture and mineralogy. Mondelli et al. (2007) point out that the resistivity values were higher in sandy layers than in clayey layers. At that time the influence of soil type on the resistivity values of the RCPTU-14 and RCPTU-15 tests were not so clear, since the resistivity value found in the RCPTU-14 test was around 50 ohm.m, while that of the RCPTU-15 test was around 20 ohm.m for saturated zone. It indicates a migration of the contamination plume through the sandy layer, based on water sampling using a direct-push sampler at depths of 8.0 to 9.0 m, since it presented low electrical resistivity, which is indicative of the presence of contaminants dissolved in the water.

Mondelli et al. (2007) conclude that the RCPTU interpretation for this particular site required collection of soil and groundwater. It was also necessary to adapt a system to measure the electrical resistivity of undisturbed soil samples in the laboratory for better understanding the resistivity variation according to the different saturation conditions and lithological characteristics that occur in tropical regions. This system, as well as the test results, is presented by Mondelli et al. (2010a). The sampling depths were selected based on the interpretation of nearby piezocone and other tests.

Mondelli et al. (2010b) present in Figure 14 the relation between the resistivity values of the saturated layers and the results of the characterization tests on soil samples collected using a direct-push sampler, in order to allow identifying and visualizing the layers susceptible to contamination. These results indicate a tendency for increasing values of resistivity with decreasing fines content, clay content or with clay mineral activity. However, layers that do not clearly follow this tendency are also shown in Figure 13, presenting low values of resistivity (between 20 and 40 ohm.m) for the sandier soils. These results demonstrate that values of resistivity ranging from 20 to 40 ohm.m for predominantly sandy layers indicate the presence of leachate in the groundwater.

This kind of interpretation goes beyond the merely comparative one of the resistivity values obtained from all the tests discussed by Mondelli et al. (2007), which had diagnosed the presence of contaminants only in tests RCPTU 14 and 15 presented in Figure 13. A certain degree of superposition of values may occur in this type of analysis, since the more clayey layers also show low values of resistivity. Since the flow of contaminants tends to occur between the sandier layers (more permeable), the problem can be simplified in an attempt to identify only these layers.

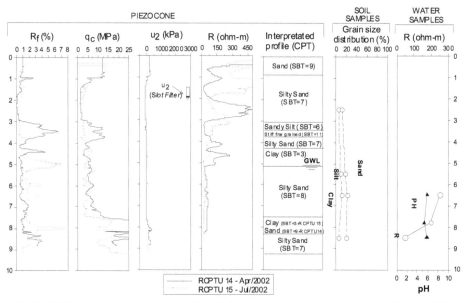

Fig. 13. RCPTU test results, grain size distribution and groundwater analyses for a MSW disposal site, adjacent to the edge of the landfill (Mondelli et al., 2007).

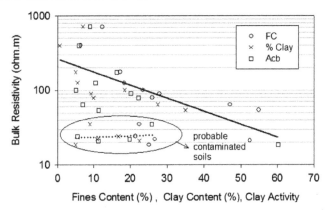

Fig. 14. Electrical resistivity in saturated soil samples with characterization of the fine fraction of soils for a MSW disposal site (Mondelli et al., 2010b).

Mondelli et al. (2010b) studies concluded that the resistivity piezocone test helped identifying contaminated zones. However, this technique presented some limitations for tropical soils, since the groundwater level is sometimes deeper than the penetrable capacity of the cone. The RCPTU tests were useful for detailing stratigraphic soil profile. The relation between corrected point resistance (q_t) and friction ration (R_f) allowed identifying the highly heterogeneous stratigraphic profiles of alluvium and colluvium. This identification for the residual soils was restricted to the behavior of the materials in response to the penetration of the piezocone, requiring soil sampling. The resistivity profiles were useful for the identification of the position of the groundwater level, and were sensitive to variations in the soil's texture and its saturation. A good relationship between the soil behavior index (Ic) and the fines content in the soil samples collected adjacent to the landfill helped the identification of contaminated regions of the saturated zone. The proper interpretation of piezocone tests enabled the identification of those layers more susceptible to contamination, which was confirmed by the chemical analyses of the groundwater samples collected from the monitoring wells. The main conclusion of this study is that the interpretation of RCPTU tests is not as straightforward as it is for sedimentary soils because the tropical soils genesis complexity.

6. Laboratory tests for geo-environmental site characterization

Laboratory tests can improve the interpretation of the in-situ test results, complementing, controlling and detailing the geo-environmental site investigation. A complete geo-environmental site investigation using current engineering practice is usually difficult to carry out because it is expensive and normally requires authorization and multidisciplinary teams. In addition, an appropriate testing campaign is time consuming and only a few private companies can do it. Limited publications on monitoring and modeling of the contamination plume caused by MSW disposal sites are available. The most commonly known case histories (Cherry et al., 1983; MacFarlane et al., 1983; Mackay et al., 1986; Kjeldsen et al., 1998; Zuquette et al., 2005) have installed more than a hundred monitoring wells, with multilevel soil and groundwater sampling, and are taken from the same sites which have been studied for years. All these studies show that geophysical tests were carried out before the intrusive tests, which were also performed to confirm contamination.

Christensen et al. (2000) explain that, contrary to the specific and known sources of contamination, – such as leaking fuel tanks (hydrocarbons) and chlorinated solvents – the municipal solid waste disposal sites are large, heterogeneous and receive various types of waste over time, have different flow pathways, hydraulic gradients and contamination plumes. Moreover, the size of the landfills will often prohibit removal of the waste and source of the leachate, suggesting that the landfill body and the leachate plume should be considered as a continuum in the context of natural attenuation.

In Brazil, most of the laboratory tests for sanitary landfills are preventive, for bottom or cover liner design, using low permeability compacted tropical soils with or without flexible membrane liners (Ritter, 1998; Boscov et al., 2001; Leite & Paraguassú, 2002; Stuermer, 2005; Azevedo et al., 2006). This approach does not consider the older waste deposits, where the contaminants are in direct contact with the natural subsoil. In Brazil, there are some researches on geo-environmental site investigation of industrial and municipal solid wastes disposal sites, using geophysics or boreholes to detect contamination plumes (Grazinolli et al., 1999; Elis & Zuquette, 2002; Lago et al., 2003; Anirban et al., 2004; Bolinelli Jr., 2004; Porsani et al., 2004;).

The in-situ techniques permit assessing the natural conditions of the soils, while the laboratory tests provide greater detail, but in minor scale, trying to represent the field conditions based on soil and water/gas samplings. Those difficulties in defining a typical soil profile, hydrogeological parameters and background resistivity (or conductivity) values due to geological complexity (like tropical sites, for example) demand more and varied site investigation techniques. After a better understanding of the site history, topography, hydrogeology and the landfill construction characteristics, laboratory tests can be carried out using disturbed and undisturbed soil samples, in order to investigate a new site for the installation of a sanitary landfill, or to confirm a pre-indicated contamination or to provide support for future remediation actions.

The typical soil profile, hydrogeological characteristics and background values can be confirmed and obtained also using laboratory tests. The soil can be characterized using geotechnical, geochemical and mineralogical techniques. Batch sorption and column leaching tests can be carried out using those geomaterials found at the study site. The landfill's leachate can be used to estimate the dispersion and retardation coefficients, for future contamination flow transport modeling. Electrical resistivity values and mineralogical constituents of non-contaminated soils can be determined in the laboratory.

6.1 Soil characterization

The characterization tests are fundamental for the classification of soils, due to their complex geological formation, and even more when performing a geo-environmental site investigation, where the interaction between the contaminants and the physical environment usually occurs. The more usual laboratory characterization tests to investigate a municipal solid waste disposal sites are:

- *Geotechnical Properties* (Lambe & Whitman, 1969): classical laboratory tests for physical soil characterization are essential for geo-environmental site investigation. Indexes like in-situ moisture content (w), natural unit weight (γ), specific gravity (Gs), grain size distribution with and without sodium hexametaphosphate for determination of clay, silt, sand and gravel contents, liquid limit (LL) and plastic limit (PL), shall be determined using disturbed and undisturbed soil samples. Also, permeability / hydraulic conductivity (K) tests are necessary for a good monitoring and pollution prevention of a waste disposal site or for future sanitary landfill, respectively. Depending on the groundwater position and the characteristics of the site, tests to determine the soil retention curve can also be important in a site investigation and monitoring program.

- *Organic Matter Content (OMC):* OMC is considered an important property of the soil solid phase, as it is responsible for retaining a good part of the organic contaminants and leachate ions, together with clay minerals and hydroxides. OMC can be determined by incineration of the soil sample in a muffle, heating to temperature of 440°C, as described in ABNT NBR 13600 (1996).

- *Blue Methyl Adsorption:* this test allows the estimative of the cation exchange capacity (CEC), specific soil surface (SE) and activity of clay minerals. This test also indicates the predominant type of clay mineral matrix in fines content of the soil sample. Lan (1977) and Pejon (1992) describe this test method in detail.

- *X-Ray Diffraction (XRD), Differential Thermal Analysis (DTA) and Gravimetrical Thermal Analysis (GTA):* These tests allow characterization of the soil fines content, by heating up the soil sample to 1000°C, in order to identify and quantify different clay minerals, hydroxides and any solid constituent of the soil. Consequently, the tests can be used to identify the soil mineralogy and its geochemical interaction with the landfill, or for assessing natural attenuation of the contaminants.

- *pH:* Determination of pH of granular materials in deionized water (usually ratio of 1 : 2.5), by measurement of the effective electrochemical concentration of H^+ ions in soil solution or waste using a combined electrode immersed in suspension. The test method is that one recommended by US EPA (1993).

- *X-Ray Fluorescence (XRF):* XRF spectrometry is used to identify elements in a soil and quantify the amount of those elements (metals) present. An element is identified by its characteristic X-ray emission wavelength (λ) or energy (E). The amount of an element present is quantified by measuring the intensity of its characteristic line. It is known that original rock composition can or cannot present natural high concentrations of some metals. For geo-environmental purposes, these conditions cannot exceed safety, health or potability exigencies for the local population. Therefore, it is noted that the levels of these metals in soils and sediments depend on the composition and proportion of them in their original solid phase. The background values need to be diagnosed before or after the waste disposal. Innov-X Systems Alpha Series (2007) presents handheld EDXRF spectrometers, ideally suited for field and laboratory analysis of soil and waste metals.

As example, Figure 15 presents a typical soil profile, characterized for construction of a future waste disposal site for a very small city in the interior of the São Paulo State, Brazil. IPT (2011) carried out a low cost geo-environmental site investigation campaign with the objective of detecting possible contamination caused by an older controlled dump near to the interest site.

The results presented in Figure 15 shows the log of the number of blows from Standard Penetration Test (N_{SPT}) up to 16.5 m depth, when the impenetrable layer was reached. During the SPT tests, deformed soil samples were collected, in order to maintain the natural moisture, and were transported to the laboratory. In the laboratory soil samples were identified based on tactile visual inspection. So, "new"soil samples were composed using the same lythology. The collected soil samples were characterized, determining moisture content (w), organic matter content (OMC), pH, electrical conductivity, grain size distribution and XRF analysis using an Innov-X handheld EDXRF spectrometer.

Figure 15 shows that the soil profile is formed by a very loose to loose sand, up to about 9.5 m depth. Between 9.5 and 13.0 m the subsoil has a medium compacity ($9 < N_{SPT} < 18$), when a layer of the same grain size is reached, but coarser and more compacted, presenting $N_{SPT} > 20$. Residual soil from sandstone was found below 15.0 m depth. The moisture content was around 7-8 %, which changed just when the groundwater level was reached, at 13.0 m depth, increasing to 17 %.

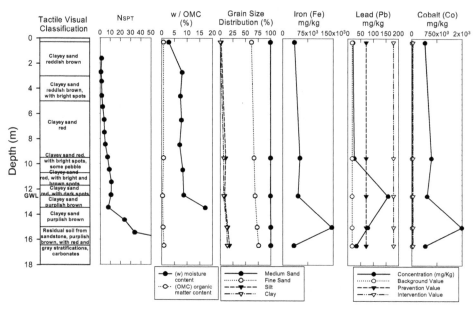

Fig. 15. Soil characterization profiles for a geo-environmental site investigation to install future sanitary trenches in a small city from Brazil.

The local soil is basically constituted of a fine to medium sand. Clay content tends to increase with depth, ranging from 0 to 16%. The organic matter content was found to be very low, ranging between 0.7 and 0.9%, tending to decrease to 0.3 to 0.4%, for the residual soil from sandstone.

The geochemical profiles obtained for iron, cobalt and lead are also presented in Figure 15. They have to be compared with the background, prevention and intervention values determined for agricultural use by the Environmental Company from São Paulo State (CETESB, 2005). There is no reference value published for iron in the São Paulo State, Brazil. It is observed that all parameters have reached peak concentrations when residual soil from sandstone is encountered. The cobalt value is greater than the intervention value and lead levels are greater than the reference value throughout all the subsoil profiling. The lead level is also greater than the intervention value for the residual soil. These high levels of Co, Pb and Fe may be associated with the constitution of the natural bedrock and in the case of where iron, iron oxides and hydroxides appear, they work like a cementing constituent for the quartz particles. The electrical conductivity (EC) values measured for the local soil were very low, ranging between 10.4 and 18.0 μS/cm, indicating that the soil is in its natural state, without ever coming into contact with leachate (highly conductive) from the older existing waste dump.

6.2 Electrical resistivity

In order to assess the changes in the resistivity values with the degree of saturation, contamination and mineralogy of some typical geomaterials found at waste disposal sites, undisturbed soil samples and leachate can be carefully selected and sampled around a particular site, for a controlled and detailed study in the laboratory, in order to interpret the in-situ resistivity maps and profiles obtained during geophysical campaigns.

Mondelli et al. (2010a) present laboratory electrical resistivity measurements to detect the contamination plumes surrounding and below a waste deposit. These authors measured electrical resistivity (R) of natural undisturbed samples as well as specimens saturated with distilled water, salt solution and original leachate from a MSW disposal site. While the specimens were naturally air dried during a specific period, the resistivity and the degree of saturation were measured. The system consists of an insulating material (PVC) mold, with two stainless-steel cylindrical electrodes, one on the top and another on the bottom of the specimen for electrical current measurement, similar to the one developed by Daniel (1997). Two more 3 mm diameter stainless-steel electrodes were attached into the middle of the specimen, observing the same distance between them (Wenner array) for differential potential transfer. For data acquisition, a Syscal Pro equipment was used. Figure 16 presents the apparatus used.

Fig. 16. Instrumented undisturbed specimen and Syscal Pro equipment used for laboratory electrical resistivity measurements (Mondelli et al., 2010a).

Mondelli et al. (2010a) concluded that the laboratory test results supported the interpretation of the in-situ test data, identifying the contamination spots, plume depths, capillarity and saturation zones and the different types of local soil. The results demonstrated the high influence of mineralogy and degree of weathering of the tropical soils on electrical resistivity values, also suggesting reference values for the study site. Figure 17 presents the results obtained for samples percolated with leachate (EC = 25,000 μS/cm and R = 0.4 ohm.m). At the beginning of the test, the resistivity values did not exceed 10 ohm.m until a 50 % degree of saturation was achieved. Below this degree of saturation, resistivity increased exponentially.

Due to the high electrical conductivity of the leachate, the resistivity values became very low, with no influence of mineralogy in this case. These results could be interpreted using Archie's Law (Archie, 1942), as studied by Mondelli (2008).

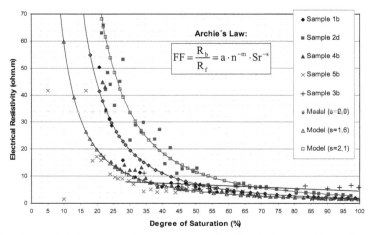

Fig. 17. Electrical resistivity versus degree of saturation for different soil samples percolated with the leachate from a MSW disposal site from Brazil (Mondelli, 2008).

6.3 Pollutant transport parameters

Estimates of pollutant transport parameters based on laboratory tests such as column, diffusion and batch equilibrium tests are also important in the more detailed phases of geo-environmental investigations (Freeze & Cherry, 1979; Rowe et al., 1988; Barone et al., 1989; Shackelford & Daniel, 1991; Yong et al., 1992; Shackelford, 1993; ong, 2001). These laboratory tests can be conducted with the objective of a better interpretation of the in-situ test results, and for a numerical modeling performance, using a good conceptual model definition for the problem. Compacted or undisturbed soil samples can be used to assess the pollutant transport parameters, targeting the contamination plume behavior when in contact with the natural environment, using equipment and methods which allows the study of the anisotropy, degree of saturation and chemical extraction.

When the determination of pollutant transport parameters is being planned, one of the most important aspects to be defined is the interest pollutants or solutions to be used during the laboratory tests. For geo-environmental investigation of MSW disposal sites, original or artificial leachate can be used. The detection of a concentration of several constituents of leachate (like heavy metals, metals, chloride, organic compounds, chemical oxygen demand and biological oxygen demand), solutions and effluents collected during the tests, need to be chemically analyzed, using emission or atomic absorption spectrometers, flame photometry, chromatography and any other appropriate techniques. The preparation and preservation proceeds to detect any kind of chemical compound need to follow the standards described in APHA, AWWA, WEF (1995).

The laboratory apparatus for column tests also needs to be planned, depending on the sample state (compacted or undisturbed), the hydraulic gradients involved defined in accordance with the field conditions and if the column system will use rigid or flexible wall permeameters. All components of the apparatus were made of non-reactive materials. This test estimates the hydraulic conductivity (k), hydrodynamic dispersion coefficient (D_{hl}) and retardation factor (R_d), due to pollutant solutions or leachate percolation through the soil sample. Once the breakthrough curves (relation between solute concentration and initial concentration – C/C_0 in function of the soil pore volumes percolated with the pollutant solution – T) are obtained for each interest solute, considering predominantly advective transport, the hydrodynamic dispersion coefficient (D_{hl}) and the retardation factor (R_d) can be assessed using the following equation proposed by Shackelford (1994):

$$\frac{C}{C_0} = \frac{1}{2}\left\{ erfc\left[\frac{1-(T/R_d)}{2\sqrt{\frac{(t/R_d)}{(uL)/D_h}}}\right] + exp\left[\frac{vL}{D_h}\right]erfc\left[\frac{1+(T/R_d)}{2\sqrt{\frac{(t/R_d)}{(uL)/D_h}}}\right]\right\} \qquad (4)$$

where: t = test time; T = time factor or soil pore volumes percolated with the pollutant solution; u = Darcy'Law specific velocity; L = height of the specimen; (uL/D_h) = Peclet number.

For non-misceble and reactive solutes, the retardation factor (R_d) can be obtained from batch sorption tests. This test can be carried out following ASTM D 4646-03 and ASTM C 1733-10, standards, using a previously defined soil-solution ratio, as 1:4 (25 g of dry soil sieved through a 40 mesh to 100 mL of leachate or pollutant solution). R_d is given by equation 5:

$$R_d = 1 + \frac{\rho_d}{n} \cdot K_d \qquad (5)$$

where: ρ_d = soil specific dried mass; n = soil porosity; K_d = partition coefficient.

Partition coefficient (K_d) represents the angular coefficient of the linear sorption isotherm, defined for an interest solute, during batch sorption tests. For the construction of the sorption isotherms, the amount of soil or solute concentration in the solution can change. When original leachate is used during the test, the solute concentrations remain constant. After a previously determined equilibrium time, the supernatants are filtered, adequatly preserved and stored at 4°C until the chemical analysis, for a limited time, depending on the interest solute. The results are expressed as the adsorption degree, as a function of the concentration. The adsorption degree (S) is defined by the following equation:

$$S = \frac{(C_0 - C_e) \cdot V}{M} \qquad (6)$$

where: C_0 = initial concentration of the solution; C_e = equilibrium concentration; V = volume of the solution in the flask; M = dry soil mass.

Frempong & Yanful (2008) and Mondelli (2008) studied the sorption capacity of tropical soils from Ghana (West Africa) and São Paulo State (Brazil), using original leachate from local MSW disposal sites, and batch and column tests. Both these studies showed the following ion sorption selectivity order: $Ca^{2+} < Cl^- < Na^+ \leq Br^- < Zn^{2+} < Fe^2 < K^+$. A significant finding from these studies is the observation that kaolinite and aluminum and iron oxyhydroxides with variable particle surface charges present in the soil allowed sorption of anions, such as Cl- and Br-, generally considered conservative nonreactive, based on leachate-liner compatibility studies on soils from temperate regions.

Frempong & Yanful (2008) found R_d ranging from 1.1 to 47.9 for Na, K, Br and Cl for clayey tropical soils. Mondelli (2008) found major values of R_d, ranging from 1 to 383 for Na, K, Fe and Zn for sandy tropical soils, with 20 % of clay content. Ritter (1998) classified the solute as essentially immobile when K_d is higher than 10. On the other hand, R_d values display the same order of magnitude as those obtained by Nascentes (2003) with similar soil and range of tested concentrations. Azevedo et al. (2003) point out that the high R_d values obtained by Nascentes (2003) can be the result of the low concentrations of metals used in the pollutant solutions. Figure 19 presents the sorption isotherm obtained for potassium (K) by Mondelli (2008), indicating that adsorption degree (S) increases along with the increase in equilibrium concentration (C_e). In this case, the linear isotherm was the one that best described the results, presenting a better fit to the testing data and K_d value of 29.8 mL/g.

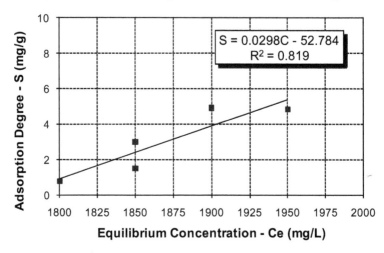

Fig. 18. Linear sorption isotherm obtained for potassium from leachate and a soil sample taken from a MSW disposal site in Brazil (Mondelli, 2008)

7. Conclusion

The chapter shortly presented and discussed the steps of a geo-environmental site investigation program for municipal solid waste disposal sites, and also, different and modern in-situ and laboratory test techniques for this purpose. A literature research was conducted on different areas of knowledge, comprising the current problems on the inadequate disposal of waste, environmental liabilities and techniques for the investigation of contaminated sites. Research results on Environmental Geotechnics were also included, aimed at the appropriate monitoring and proper installation of sanitary landfills. Therefore, it can be concluded that it is fundamental to know exactly all the goals, local interests, background factors and resources involved in each case, for an effective and optimized geo-environmental site investigation. This approach will provide positive results for the local population and for the physical environment.

8. Acknowledgments

The authors gratefully acknowledge the Brazilian research agencies FAPESP (State of São Paulo Research Foundation), CAPES (Brazilian Federal Agency for Support and Evaluation of Graduate Education) and CNPq (National Council for Scientific and Technological Development) for funding their researches.

9. References

ABNT NBR – 13600 (1996). *Solo – Determinação do Teor de Matéria Orgânica por Queima a 440°C.* Brazilian Association for Technical Standards, São Paulo-SP, Brazil.

APHA, AWWA, WEF (1995). *Standard Methods for the Examination of Water and Wastewater.* American Public Health Association, American Water Works Association, Water Environmental Federation. Managing Editor Mary Ann H. Franson. 19th Edition, U. S. A., pp. 3-1 – 3-2.

ASTM D 5778-07. *Standard Test Method for Electronic Friction Cone and Piezocone Penetration Testing of Soils.* ASTM International, Pennsylvania, U.S.A.

ASTM D 4646- 03 (Reapproved 2008). *Standard Test Method for 24-h Batch-Type Measurement of Contaminant Sorption by Soils and Sediments.* ASTM Intern., Pennsylvania, U.S.A.

ASTM C 1733-10. *Standard Test Method for Distribution Coefficients of Inorganic Species by the Batch Method.* ASTM Intern., Pennsylvania, U.S.A.

Anirban, D. E.; Matasovic, N. & Dunn, R. J. (2004). Site Characterization of Five Hazardous Waste Landfills. *Proceedings of the 2nd International Conference on Site Characterization,* Porto, Portugal, September 2004, pp. 1075-1080.

Archie, G. E. (1942). The Electrical Resistivity Log as an Aid in Determining Some Reservoir Characteristics. *Transactions of American Institute of Mining and Metallurgy Engineering,* Vol.146, pp. 54-62.

Azevedo, I. C. D.; Nascentes, C. R.; Matos, A. T. & Azevedo, R. F. (2006). Determination of Transport Parameters for Heavy Metal in Residual Compacted Soil Using Two Methodologies. *Canadian Journal of Civil Engineering,* Vol.33, pp. 912-917.

Azevedo, I. C. D. , Nascentes, C. R.; Azevedo, R. F.; Matos, A. T. & Guimarães, L. M. (2003). Coeficiente de Dispersão Hidrodinâmica e Fator de Retardamento de Metais Pesados em Solo Residual Compactado. *Solos e Rochas,* Vol.26, pp.229-249.

Barone, F. S.; Yanful, E. K.; Quigley, R. M. & Rowe, R. K. (1989). Effect of Multiple Contaminant Migration on Diffusion and Adsorption of Some Domestic Waste Contaminants in a Natural Clayey Soil. *Canadian Geotechnical Journal,* Vol.26, pp. 189-198.

Bolinelli Jr., H. L. (2004). *Piezocone de Resistividade: Primeiros Resultados de Investigação Geoambiental em Solos Tropicais.* MSc, Department of Civil Engineering, São Paulo State University, Bauru-SP, Brazil.

Boscov, M. E. G.; Cunha, I. I. & Saito, R. (2001). Radium Migration Through Clay Liners at Waste Disposal Sites. *Science of t he Total Environment,* Vol.266, pp. 259-264.

Campanella, R. G. (2008). Geo-Environmental Site Characterization. *Proceedings of the 3rd International Conference on Geotechnical and Geophysical Site Characterization,* Taipei – Taiwan, Taylor & Francis Group, pp. 3-16.

CETESB (2005). *Relatório de Estabelecimento de Valores Orientadores para Solos e Águas Subterrâneas no Estado de São Paulo.* São Paulo-SP, Brasil.

Cherry, J. A.; Gillham, R. W.; Anderson, E. G. & Johnson, P. E. (1983). Migration of Contaminants in Groundwater at a Landfill: a Case Study 2. Groundwater Monitoring Devices. *Journal of Hydrology*, Vol.63, pp. 31-49.

Christensen, T. H., Bjerg, P. L. & Kjeldsen, P. (2000). Natural attenuation: a feasible approach to remediation of ground water pollution at landfills? *Ground Water Monitoring & Remediation*, Vol.20, pp. 69-77.

Dahlin, T.; Bernstone, C. & Loke, M. H. (2002). A 3-D Resistivity Investigation of a Contaminated Site at Lernacken, Sweden. *Geophysics*, Vol.67, pp. 1692-1700.

Daniel, C.R. (1997). *An Investigation of the Factors Affecting Bulk Soil Electrical Resistivity*. BASc, University of British Columbia, Department of Civil Engineering.

Davies, M. P. & Campanella, R. G. (1995). Environmental Site Characterization Using In-Situ Testing Methods, *48th Canadian Geotechnical Conference*, Vancouver-BC, Canada.

Elis, V. R. & Zuquette, L. V. (2002). Caracterização Geofísica de Áreas Utilizadas para Disposição de Resíduos Sólidos Urbanos. *Revista Brasileira de Geociências*, Vol.32, pp. 119-134.

Frempong, E. M. & Yanful, E. K. (2008). Interactions Between Three Tropical Soils and Municipal Solid Waste Landfill Leachate. *Journal of Geotechnical and Geoenvironmental Engineering*, Vol.134, pp.379-396.

Freeze, R. A. & Cherry, J. A. (1979). *Groundwater*. Prentice Hall, Inc. Englewood Cliffs, New Jersey, U. S. A.

Geoprobe Direct Image (2004). *MP 6500 – Membrane Interface Probe (MIP): User Manual*. Document N. 3083, Revision 1.0 10-12-2004, Kejr, Inc., Salina, Kansas, U.S.A.

Grazinolli, P. L.; Costa, A.; Campos, T. M. P. & Vargas Jr., E. A. (1999). Aplicações do Radar de Penetração no Solo (GPR) e da Eletrorresistividade para a Detecção de Compostos Orgânicos. *Proceeding of the 4º Congresso Brasileiro de Geotecnia Ambiental*, São José dos Campos, Brazil, December 1999, pp 127-13.

Innov-X Systems Alpha Series (2007). *X-Ray Fluorescence Spectrometers*. P/N 100392, Revision B. Woburn, MA, U. S. A.

IPT (2011). *Estudo de Viabilidade para Ampliação da Área de Disposição Final de Resíduos Sólidos Urbanos no Município de Anhembi-SP*. Institute for Technological Research, Center for Energy and Environmental Technologies. Parecer Técnico 19455-301/11, complementar ao Parecer Técnico 19287-301/11.

Kjeldsen, P.; Bjerg, P. L.; Rügge, K.; Christensen, T. H. & Pedersen, J. K. (1998). Characterization of an Old Municipal Landfill (Grindsted, Denmark) as a Groundwater Pollution Source: Landfill Hydrology and Leachate Migration. *Waste Management & Research*, Vol.16, pp. 14-22.

Lago, A. L.; Silva, E. M. A.; Elis, V. R. & Giacheti, H. L. (2003). Aplicação de Eletrorresistividade e Polarização Induzida em Área de Disposição de Resíduos Sólidos Urbanos em Bauru-SP. *Proceedings of the 8th International Congress of the Brazilian Geophysical Society*, Rio de Janeiro, Brazil, November 2003, CD-ROM.

Lambe, T. W. & Whitman, R. V. (1969). *Soil Mechanics*. New York: J. Wiley, 553p.

Lan, T. N. (1977). Um Nouvel Esai D'identification des Sols: L'essai au Bleu de Methyléne. *Bulletin Liaison Laboratoire dês Ponts et Chaussée*, Paris, pp.136-137.

Larsson, R. (1995). Use of a Thin Slot as Filter in Piezocone Tests. *Proceedings of International Symposium on Cone Penetration Testing (CPT'95)*, Linköping, Sweden, pp. 35-40.

Leite, A. L. & Paraguassú, A. B. (2002). Diffusion of Inorganic Chemicals in Some Compacted Tropical Soils. *Proceedings of the 4th International Congress In Environmental Geotechnics*, Rio de Janeiro, Brazil, August 2002, pp 39-45.

Loke, M. H. & Barker, R. D. (1996). Practical Techniques for 3D Resistivity Survey and Data Inversion. *Geophysical Prospecting*, Vol.44, pp. 499-523.

Lunne, T.; Robertson, P. K. & Powell, J. (1997). *Cone Penetration Test in Geotechnical Practice*. Blackie Academic & Professional, London.

Mackay, D. M.; Freyberg, D. L.; Roberts, P. V. & Cherry, J. A. (1986). A Natural Gradient Experiment on Solute Transport in a Sand Aquifer: 1. Approach and Overview of Plume Movement. *Water Resources Research*, Vol.22, pp. 2017-2029.

MacFarlane, D. S.; Cherry, J. A.; Ghillham, R. W. & Sudicky, E. A. (1983). Migration of Contaminants in Groundwater at a Landfill: a Case Study, 1. Groundwater Flow and Plume Delineation. *Journal of Hydrology*, Vol.63, pp. 1-29.

McNeill, J. D. (1980). Electromagnetic Terrain Conductivity Measurement at Low Induction Numbers. *Geonics Technical Note*, No.6, 15p.

Mondelli, G. (2008). *Integração de Diferentes Técnicas de Investigação para Avaliação da Poluição e Contaminação de uma Área de Disposição de Resíduos Sólidos Urbanos*. Ph.D. Thesis, Department of Geotechnical Engineering, University of São Paulo, São Carlos-SP, Brazil.

Mondelli, G.; Giacheti, H. L.; Boscov, M. E. G.; Elis, V. R. & Hamada, J. (2007). Geoenvironmental Site Investigation Using Different Techniques in a Municipal Solid Waste Disposal Site in Brazil. *Environmental Geology*, Vol.52, pp. 871-887.

Mondelli, G.; Giacheti, H. L. & Elis, V. R. (2010a). The Use of Resistivity for Detecting MSW Contamination Plumes in a Tropical Soil Site. *Proceedings of 6th International Conference on Environmental Geotechnics*, New Delhi, India, pp. 1544-1549.

Mondelli, G.; Giacheti, H. L. & Howie, J. A. (2010b). Interpretation of Resistivity Piezocone Tests in a Contaminated Municipal Solid Waste Disposal Site. *Geotechnical Testing Journal*, Vol. 33, N. 2, pp. 123-136.

Monteiro Santos, F. A. (2004). 1-D Laterally Constrained Inversion of EM34 Profiling Data. *Journal of Applied Geophysics*, Vol.56, pp. 123-134.

Nascentes, C. R. (2003). *Coeficiente de Dispersão Hidrodinâmica e Fator de Retardamento de Metais Pesados em Solo Residual Compactado*. MSc, Department of Civil Engineering, Federal University of Viçosa, Viçosa-MG, Brazil.

Pejon, O. J. (1992). *Mapeamento Geotécnico da Folha de Piracicaba-SP (Escala 1:100.000): Estudo de Aspectos Metodológicos, de Caracterização e de Apresentação dos Atributos*. Ph.D. Thesis, Department of Geotechnical Engineering, University of São Paulo, São Carlos-SP, Brazil.

Porsani, J. L.; Malagutti Filho, W.; Elis, V. R.; Fisseha, S., Dourado, J. C. & Moura, H. P. (2004). The Use of GPR and VES in Delineating a Contamination Plume in a Landfill Site: A Case Study in SE Brazil. *Journal of Applied Geophysics*, Vol.55, pp. 199-209.

Ritter, E. (1998). *Efeito da Salinidade na Difusão e Sorção de Alguns Íons Inorgânicos em um Solo Argiloso Saturado*. PhD, Department of Civil Engineering, Federal University of Rio de Janeiro, Rio de Janeiro-RJ, Brazil.

Robertson, P. K. (1998). Geo-Environmental Investigation, Characterization and Monitoring Using Penetration Techniques. *Simpósio Brasileiro de Geotecnia Ambiental*, São Paulo-SP, Brazil.

Robertson, P. K.; Campanella, R. G., Gillespie, D. & Greig, J. (1986). Use of piezometer cone data, *Proc. of the In-Situ-86, ASCE Specialty Conference*, p. 1263-1280

Robertson, P. K. & Cabal, K. L. (2008). *Guide to Cone Penetration Testing for Geo-Environmental Engineering*. Gregg Drilling & Testing, Inc., 2nd Edition, 84 p.

Robertson, P. K. & Campanella, R. G. (1988). *Guidelines for Geotechnical Design Using CPT and CPTU Data*. Report FHWA, 340p.

Rowe, R. K.; Caers, C. J. & Barone, F. S. (1988). Laboratory Determination of Diffusion and Distribution Coefficients of Contaminants Using Undisturbed Clayey Soil. *Canadian Geotechnical Journal*, Vol.25, No.1, pp. 108-118.

Shackelford, C. D. (1993). Contaminant Transport. *Geotechnical Practice for Waste Disposal*, Chapman & Hall, London, pp. 33-65.

Shackelford, C. D. (1994). Critical Concepts for Column Testing. *Journal of Geotechnical Engineering*, Vol.120, pp.1804-1828.

Shackelford, C. D. & Daniel, D. E. (1991). Diffusion in Saturated Soil. I: Background. *Journal of Geotechnical Engineering*, Vol.117, No.3, pp. 467-484.

Shinn II, J.D. & Bratton, W.L. (1995). Innovations with CPT for environmental site characterization", *Proceedings of CPT'95*, Vol. 2, pp. 93-98.

Stuermer, M. M. (2005). *Contribuição ao Estudo de um Solo Saprolítico como Revestimento Impermeabilizante de Fundo de Aterros de Resíduos*. PhD, Department of Structures and Geotechnical Engineering, University of São Paulo, São Paulo-SP, Brazil.

Telford, W. M.; Geldart, L. P.; Sheriff, R. E. & Keys, D. A. (1990). *Applied Geophysics*, Cambridge University Press, 860 p.

Thornthwaite, C.W. & Mather, J.R. (1955). *The water balance*. Publications in Climatology. New Jersey: Drexel Institute of Technology, 104p.

US EPA (1989). Seminar on Site Characterization for Subsurface Remediations. *United States Environmental Protection Agency*, Technology Transfer, Report CERI-89-224, 350p.

US EPA (1993). *SW-846 pH in Liquid and Soil*. Method 9040 (Liquid) and SW-846 Method 9045 (Soil).

Ustra, A. T.; Elis, V. R.; Mondelli, G.; Zuquette, L. V. & Giacheti, H. L. (2011). Case Study: A 3D Resistivity and Induced Polarization Imaging from Downstream a Waste Disposal Site in Brazil. *Environmental Earth Sciences*, in press (online).

Weemes, I. (1990). *A Resistivity Cone Penetrometer for Ground-Water Studies*. MASc, University of British Columbia, Department of Civil Engineering.

Yong, R. N. (2001). *Geoenvironmental Engineering: Contaminated Soils, Pollutant Fate & Mitigation*. CRC Press, USA, 307p

Yong, R. N.; Mohamed, A. M. O. & Warkentin, B. P. (1992). *Principles of Contaminant Transport in Soils*. Elsevier Science Publishers B.V., 327 p.

Zuquette, L. V.; Palma, J. B. & Pejon, O. J. (2005). Environmental Assessment of an Uncontrolled Sanitary Landfill. *Bulletin of Engineering Geology and the Environment*, Vol.64, pp.257-271.

2

Application of Geophysical Methods to Waste Disposal Studies

Cristina Pomposiello, Cristina Dapeña, Alicia Favetto and Pamela Boujon
Instituto de Geocronología y Geología Isotópica (INGEIS, CONICET-UBA)
Argentina

1. Introduction

Geophysical methods provide information on the distribution of certain physical parameters in the sub-surface, which can be linked to the direct observations. Thus it is called an indirect observation method and it does not provide a "photo" of the sub-surface but it suggests a model of the underground derived from interpreting the distribution of these physical parameters. Clay and granite, for example, have different densities, acoustic velocities, elastic parameters, electrical conductivities, magnetic susceptibilities, and dielectric constants. So, geophysical methods are designed to exploit some of the physical properties of a target feature that is in contrast with its host environment, e.g., the low density nature of a void is in contrast to the high density nature of surrounding bedrock, etc. Geophysics should never be a stand-alone tool, but complementary to direct observations, which provide geological/hydrogeological background information (such as some of those methods seen in Table 1).

There are two general types of geophysical methods: 1) active, which measure the sub-surface response to electromagnetic, electrical, and seismic energy generated by artificial sources; and 2) passive, which measure the earth's ambient magnetic, electrical, and gravitational fields. Geophysical instruments are designed to map spatial variations in the physical properties of the Earth. A gravimeter, for example, is designed to measure spatial variations in the strength of Earth's gravitational field.

Sanitary landfill is the most common way to eliminate solid urban wastes. They have a heterogeneous structure due to random origin of the disposed waste. Geophysical methods are particularly valuable because they are non-destructive and non-invasive. An important problem associated with this practice is leachate production and the related groundwater contamination. Leachate electrical conductivity is often much higher than that of natural groundwater and it is this large contrast that enables contamination plumes to be detected using geophysical methods.

Ground Penetrating Radar (GPR) and Electrical methods such as Electrical Tomography (ET) and Vertical Electrical Sounding (VES) have been found to be especially useful for these kinds of environmental studies, due to the conductive nature of most contaminants, and they can be important tools for the detection and mapping of landfills, trenches, buried wastes and drums, or other underground structures.

Geophysical Method	Measured Parameter	Physical Property	Physical Property Model	Applications
Electrical resistivity	Potential difference and induce current	Electrical resistivity	Electrical resistivity model	Determine depth and thickness of geological layers.
Induced polarization (IP)	Polarization voltages or frequency dependent ground resistance	Electrical capacity	Electrical capacity model	Determine electrical conductive targets such as clay content (or metallic)
Ground penetrating radar (GPR)	Travel times and amplitudes of EM waves.	Dielectric constant, magnetic permeability and electrical conductivity.	EM velocity model	Detect both metallic and non-metallic targets
Magnetics	Spatial variations in the strength of magnetic field of the Earth	Magnetic susceptibility and remanent magnetization	Model depicting spatial variations in the magnetic susceptibility of subsurface.	Map geological structures and detect buried drums, tanks, and other metal objects
Gravity	Spatial variations in the strength of gravitational field of the Earth	Bulk density	Model depicting spatial variations in the density of sub-surface	Determine any geologic structure involving mass variations.

Table 1. Geophysical methods employed for environmental investigations.

1.1 Sub-surface electrical resistivity

Several conduction mechanisms are possible in typical sub-surface material. They are: a) electronical conduction, b) semiconductors and c) ionic conduction in liquids.

a. Electronic conduction occurs in pure metals. In this mechanism the charge carriers are electrons and their high mobility gives a very low resistivity ($< 10^{-8}$ ohm-m).

b. Semiconduction occurs in minerals such as sulphides and typically in igneous rocks. In this case the charge carries are electrons, ions or holes. Compared to metals, the mobility and number of charge carriers are lower and thus the resistivity higher (typically 10^{-3} to 10^{-5} ohm-m).

c. Ionic conduction in liquids (or electrolytic conduction) occurs when current flows via the movement of ions in aqueous fluids or molten materials. The resistivity of rocks is greatly dependent on the degree of fracturing and the percentage of the fractures filled with fluids. Igneous and metamorphic rocks normally have higher resistivity values compared with sedimentary rocks (typically 1 to 100 ohm-m), which are usually more porous and have higher water content. The resistivity values are largely dependent on the porosity of the rocks and the salinity of the ground water.

Electrical resistivity distribution is useful for determining shallow and deep geological and hydrogeological conditions. Geoelectrical surveys are commonly used in hydrogeological, mining and geotechnical investigations. More recently, it has been used for environmental surveys (Loke, 1999). The resistivity data can be used to identify, delineate and map the sub-surface defining such things as electrical conductor contamination plumes, lithologic units with clay, the salt water/fresh water interface and the vadose zone.

The deposits in landfill sites have different origins, for example domestic and industrial wastes, soils and exhumed geological materials. They have a complex material composition; they are non-uniformly compacted and also have a non-uniform decomposition process, so their physical properties would present a wide range of variation. For example the electrical resistivity has values that range from 1.5 to 20 ohm-m in different landfills (Meju et al. 2000). Leachate is a liquid formed from decomposed waste and it can contain ground water and percolated rainwater. Inorganic pollutants increase liquid conductivity due the presence of dissolved salts. Consequently, the electrical resistivity of leachate is often very much lower than natural groundwater. For example, the fresh water has a resistivity around 200 ohm-m and the saline water has a resistivity of around 10 ohm-m. Other type of pollutants, such as organic compounds can reduce leachate conductivity.

1.2 Geoelectrical methods

The sub-surface resistivity distribution can be determined by making measurements on the ground surface. From these measurements, the true resistivity can be inferred. The ground resistivity is related to various geological parameters such as the mineral and fluid content, porosity and degree of water saturation in the rock.

The resistivity measurements are normally made using a man-made source of electrical current that is applied to the earth through grounded electrodes (C1 and C2 in Figure 1). The resulting potential field is measured along the ground using a second pair of electrodes (P1 and P2). The transmitting and receiving electrode pairs are referred to as dipoles. By varying the unit length of the dipoles as well as the distance between them, the horizontal and vertical distribution of electrical properties can be calculated. From the current (I) and voltage (V) values, the apparent resistivity (ρa) is calculated, $\rho a = k\, V\, /I$ where k is the geometrical factor which depends on the arrangement of the four electrodes.

The apparent resistivity is not the true resistivity of the surface, but this value equals that of the resistivity of homogeneous ground measured from the same electrode arrangement. From the apparent resistivity values it is possible obtain the true surface resistivity using an inversion method.

Fig. 1. A conventional four electrode array to measure the sub-surface resistivity.

1.2.1 1D Vertical Electrical Sounding (VES)

The vertical electrical sounding is performed changing the electrode spacing over a common centre point. The electrical current is applied to A and B electrodes and the potential V is measured between M and N electrodes. The VES array consists of a series of the electrode combinations AMNB with gradually increasing distances between the electrodes for subsequent combinations. The Schlumberger array is shown in Figure 2. The depth of sounding increases with the distance between A and B electrodes.

VES provides measurements of apparent resistivity obtaining a sounding curve (Figure 3a), by plotting resistivity ρa against the half distance between the current electrodes (AB/2) with k= π (AO² - MN²/4)/MN. It is interpreted quantitatively to derive thickness and resistivity of sub-surface layers using the appropriate software which determines a depth-layer model of resistivity (1-D-model). It provides the vertical resistivity variation under the centre point and it is applicable to horizontally layered homogenous ground (Figure 3b).

This method is a very important tool employed in the geophysical exploration of the shallow sub-surface. The normal depth range that is investigated is from a few centimetres to a few hundred meters. These techniques are thus particularly applicable in groundwater, and environment characterizations. The distinction between mainly clayey/silty layers (aquitards) and dominantly sandy/gravely beds (aquifers) are specially distinguishable to determine. Furthermore, it can be used to determine saltwater intrusion and contaminated plumes

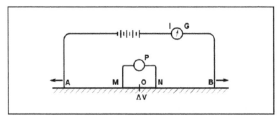

Fig. 2. Schlumberger array, depth of sounding controlled by the distance between A and B.

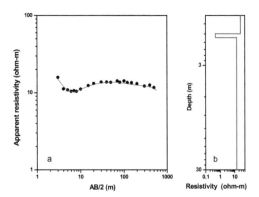

Fig. 3. a) Sounding curve, apparent resistivity vs. the half distance between the current electrodes. b) 1D model, depth vs. resistivity (log-log, representation).

1.2.2 2D Electrical Tomography (ET)

2D imaging assumes low variation of the third dimension. 3D surveys involve large amount of measurements and more data processing.

In an electrical tomography an array of regularly – spaced electrodes is deployed. They are connected to a central control unit via multi-core cables. The common arrays used are dipole-dipole, Schumberger and Wenner, depending on application and the resolution desired (Loke, 1999). The advantages and disadvantages of these arrays will be used to choose the appropriate configuration in each case. The dipole-dipole array is present in Figure 4. In this case, the spacing between the current electrodes pair, AB is given as "a" which is the same as the distance between the potential electrodes pair MN. The same process is repeated for measurements with different spacing ("2a" to "na"). The apparent resistivity is calculated with k = π(n(n+1)(n+2)a, where n is the level. The median depth of investigation of this array also depends on the "n" factor, as well as the "a".

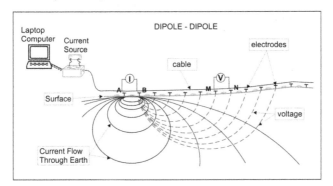

Fig. 4. Multi-channel dipole-dipole surveys.

Resistivity data are then recorded via complex combinations of current and potential electrode pairs to build up a pseudo cross-section of apparent resistivity beneath the survey profile (Figure 5).

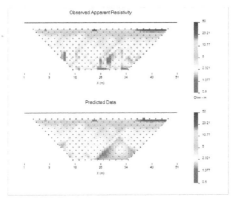

Fig. 5. An example of observed apparent resistivity (a = 2 m and 231 data) and predicted data by an inversion method.

Electrical tomography has many applications in geology, hydrogeology and environmental studies, for example:

a.   Determination of the depth and thickness of geological strata.
b.   Detection of lateral changes and locating anomalous geological conditions
c.   Locating buried wastes (e.g., locate landfill)
d.   Mapping saltwater intrusion and contaminated plumes.

This electrical resistivity imaging survey is widely used to control the depth extent and geometry of the landfill. Frequent monitoring landfill leachate with this technique could allow early leak detection.

1.2.3 GPR

Ground penetrating radar (GPR) uses pulsed high frequency radio waves (10 to 2000 MHz) to probe the sub-surface without disturbing the ground. Energy is radiated down-ward into the ground from a transmitter and is reflected back to a receiving antenna (Figure 6). The reflected signals are recorded continuously as the GPR is pulled over the ground surface and provides a real-time cross section or image of the subsurface (Figure 7). GPR can be used in a great variety of materials such as rock, soil, ice, fresh water and road surfaces.

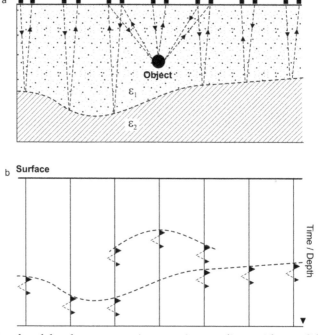

Fig. 6. a) Travel paths of the electromagnetic waves in a medium with two different dielectric proprieties and a buried object. b) Travel times of the reflections. It is a function of the layer thickness and the velocities. The isolated body presents a hyperbola.

In this method the measured parameters are the travel time and amplitudes of reflected pulsed electromagnetic energy and they depend on physical properties such as the dielectric constant and the conductivity of the material. Essentially, a reflection occurs when there is a change in the dielectric constant of materials in the sub-surface (Table 2). The dielectric constant is defined as the capacity of a material to store a charge when an electric field is applied relative to the same capacity in vacuum, and can be computed as: $\varepsilon_r = (c/v)^2$, where: ε_r = the relative dielectric constant, c = the speed of light (30 cm/nanosecond), and v = the velocity of electromagnetic energy passing through the material.

The principles involved, in this method, are similar to reflection seismology except that electromagnetic energy is used instead of acoustic energy.

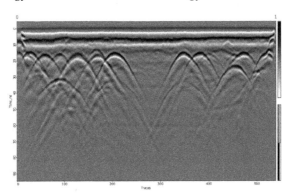

Fig. 7. Example of a GPR radargrama.

Material	Dielectric relative permittivity	Conductivity (mS/m)	Velocity (m/ns)	Attenuation (dB/m)
Air	1	0	0.3	0
Distillate water	80	0.001	0.033	0.002
Fresh water	80	0.5	0.033	0.1
Sea water	80	3000	0.01	1000
Dry sand	3-5	0.01	0.15	0.01
Wet sand	20-30	0.1-1	0.06	0.03-0.3
Limestone	4-8	0.5-2	0.12	0.4-1
Shale	5-15	1-100	0.09	1-100
Silt	5-30	1-100	0.07	1-100
Clay	5-40	2-1000	0.06	1-300
Granite	4-6	0.01-1	0.13	0.01-1
Ice	3-4	0.01	0.16	0.01

Table 2. Relative dielectric permittivity (relative dielectric constant), conductivity, velocity and attenuation for different media.

The penetration depth of GPR is determined by the antenna frequency and the electrical conductivity of the sub-surface materials being profiled (Daniels, 2004) and the choice of working frequency is dependent on the depth of penetration. As conductivity increases, the penetration depth also decreases. This is because the electromagnetic energy is more quickly converted into heat, causing a loss in signal strength at depth. Soils having high electrical conductivity rapidly attenuate radar energy, restrict penetration depths, and severely limit the effectiveness of GPR. The electrical conductivity of soils increases with increasing water, clay and soluble salt contents. Optimal depth penetration is achieved in ice where the depth of penetration can achieve several hundred metres. Good penetration is also achieved in dry sandy soils or massive dry materials such as granite, limestone, and concrete where the depth of penetration could be up to 15 metres.

The GPR has many applications in environmental studies, principally the characterization of abandoned landfills: thickness of the deposit, location of metal and non-metal objects (barrels, voids, etc.), soil profiles, groundwater level and contaminant plume.

The GPR has several advantages, for example: a) measurements are relatively easy to make, b) the lateral and vertical resolution is very high and c) the antenna may be pulled by hand or vehicle.

Due to the complexity of the fill materials of the landfill, GPR signals in landfills are very complex and difficult for interpretation, Also the principal limitation is the presence of clay-rich soil or saline groundwater which can reduce the exploration depth due to the attenuation the radar signal. However, GPR can be used as a secondary tool for some localized interested area where the GPR survey might be able to provide additional detailed information and it is easier to follow the contamination plume due to the attenuation the radar signal outside the landfill.

2. Case studies

This chapter presents the geophysical results of multidisciplinary projects at two urban solid wastes, one sites in the city of Gualeguaychú, Province of Entre Ríos (Pomposiello et al, 2009), and the other in the city of San Carlos de Bariloche, Province of Rio Negro (Pomposiello et al, 2008), Argentina. As part of an environmental project several tomographies using dipole-dipole electrode array, vertical sounding using Schlumberger electrode configurations and GPR profiles with 150 –500 MHz antennae were performed within and outside the landfills and they were interpreted using one-dimensional and bidimensional models. These studies such as electrical resistivity imaging and GPR techniques were used to locate and monitor leachate plumes in landfill sites.

2.1 Applied geophysical methods

2.1.1. Dipole-Dipole (DD)

Geoelectrical profiles were performed using the DD configuration with 21 electrodes (spaced a = 5 with n =1, ..., 18) to give a total profile length of 100 meters. The maximum depth to which the experimental data gives reliable information on the electrical resistivity of medium was estimated. The resistivity model was obtained from the data inversion using the DCINV2D algorithm of Oldemburg et al. (1993). We compared the results using half-space reference models of 0.001, 0.01 and 0.1 mS/m. The depth of investigation index (DOI) as defined by Oldenburg et al. (1994) was calculated using a reference model of 0.001 mS/m and a cut off = 0.5.

2.1.2 Vertical Electrical Sounding (VES)

For the evaluation of the electrical resistivity in the deeper layers **VES** were made using Schlumberger configuration with opening distance (AB) between 6 m and 1000 m, which allows a model to be found focused on the measuring point and a penetration of 200 m and 300 meters. Models were obtained using the inversion software IPIWIN (Bobachev et al., 2000).

2.1.3 Ground Penetrating Radar (GPR)

The GPR allowed the observation of the shallowest levels (less than 10 m) and to distinguish the boundary between the deposits of waste and the area which is in contact with leachate. The GPR profiles were performed with 500 MHz antennas (screened) and 150 MHz (air). The commercial program Prism version 2.01 by Radar System, Inc (http://www.radsys.lv, 2004) was used.

2.2 Relationship between electrical conductivity and hydrochemistry parameters

Chemical analyses of contaminated groundwater by leachate in different geographical regions demonstrated that groundwater electrical conductivity has a strong lineal correlation with total dissolved solids (TDS) and chloride content (Cl-) (Meju 2000 & his references)

The leachate composition varies with the age of the landfill, and as a result recent solid wastes have higher contents of organic acids, ammonium and total dissolved salts (TDS), accordingly their concentrations decrease when biodegradation begins. As a consequence, it is possible to estimate the age of the landfill using these measurements (Farquhar, 1989; Meju, 2000).

In addition, Meju (2000) indicates that measuring the electrical conductivity of the saturate substrate (σ_b) makes possible to predict the values of TDS applying the following equations:

$$\log \sigma_b = -0{,}3215 + 0{,}7093 \,{}^{*}\log \text{TDS} \tag{1}$$

$$\log \sigma_b = -0{,}333 + 0{,}6453 * \log \text{TDS} \tag{2}$$

where: σ_b is expressed in mS/m and TDS in mg/L. Equations (1) and (2) determine the minimum and maximum values respectively

From TDS values it is possible to predict the water electrical conductivity (σ_w) and the chloride content using the following equations

$$\log \text{TDS} = 0{,}8 + 1{,}015 * \log \sigma_w \tag{3}$$

$$\log \text{Cl-} = -0{,}256 + 1{,}2{}^{*}\log \sigma_w \tag{4}$$

where: σ_w is expressed in mS/m and the last parameters in mg/L.

These relationships were calculated in a sanitary landfill in Durban, South Africa (Bell & Jermy, 1995). Data from others landfills in Australia (Buselli et al., 1990) and Canada (Birks & Eyles, 1997) also show a similar lineal tendency to the Durban values.

The water content (W) can be calculated applying equation (1) and considering the stratum free of clay. (Yaramanci, 1994):

$$\sigma_b = \sigma_w * W^m \tag{5}$$

where: W is the water content in percentage of volume (vol %) and m is the cementation factor, which value is between 1,6 y 1,9.

Equation (2) was established for relatively homogeneous materials and it can be non valid for environments with solid wastes and significant quantities of conductors, metals and clay.

As a result, the age of landfills can be determined considering the relations and the changes in the concentration of chemical parameters of lixiviates such as Cl-, TDS, pH, between others (Farquhar, 1989; Meju, 2000).

2.3 Gualeguaychú landfill

2.3.1 Background

The study area corresponds to the actual municipal sanitary landfill of the city of Gualeguaychú in the Southern Entre Ríos Province, Argentina (Figure 8). The landfill is near the south border of the city. This city has a population of 80000, and it has numerous commercial, industrial and agricultural activities. The landfill operations were suspended eleven years ago and no environmental studies were performed to evaluate its actual state and outline the contamination zone. Currently, the operation of the waste deposit has been resumed by the Municipal Urban Hygiene Enterprise, which solicited an evaluation of the actual state of the old sanitary landfill.

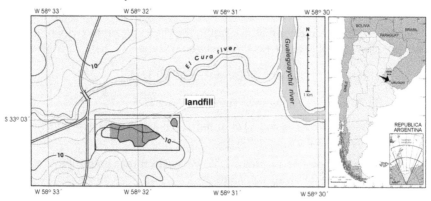

Fig. 8. Location of the Gualeguaychú landfill relevant to the Entre Rios province, Argentina.

At present, in Gualeguaychú there are two landfills overfilled and closed with different ages. The oldest and biggest was closed in 1998 and the smallest was closed in 2003. In the case of the first one, there are no records of its construction and it has neither leachate collection system nor a collection system for gases, although it is observed that the residues have been deposited in parallel cells. Now, a third landfill is in operation. This work only presents the results obtained in the first landfill.

The site of the landfill was initially used as a place of extraction of materials (ex quarry Irazusta), allowing the troughs to be used for the storage of solid waste. In particular, this site was an open dump where solid wastes were deposited and then covered by the excavated material (Figure 9). This material belongs to the Punta Gorda Group with lithology consisting of silt and clay with calcareous concretions and calcareous hardpan layers that appear at different levels (locally known as "tosca") and in the lower section the presence of clay is richer than the upper one and it also has some gypsum flake crystals. Wastes were not recycled prior to deposition and arranged in parallel modules (cells) of about 0.8 -1 m wide. Prezzi et al. (2005) estimated that the cells reach a maximum depth around 2 to 3 meters.

Fig. 9. A photo of the landfill landscape.

These landfills are located below the contour level of 10 m and the regional slope is approximately N-NE, toward the El Cura stream and the Gualeguaychú River with a very low gradient (Figure 8). The El Cura stream is the natural collector of the regional surface flow and always carries water generating a strong environmental impact due to the discharges of untreated sewage effluents of the city.

The boundaries and surface of the waste are irregular, following the geometry that comes from the alignment of cells and providing a soft and notable undulation. At the northern edge a channel is formed between the landfill and the road, where water accumulates. This deposit is divided into three parts, and between the central and west sector there is a waste-free zone which functions like a permanent course of water (stream W, Figure 10). The eastern sector is separated from the central one by a big area free of waste which also functions as an intermittent course of water (stream E, Figure 10) Both streams have a N-S direction carrying plenty of water after large precipitations. They discharge in an undefined way at the northern edge on the other side of the road forming flooded areas (Figure 10). The depressions between cells act as water reservoirs after rains.

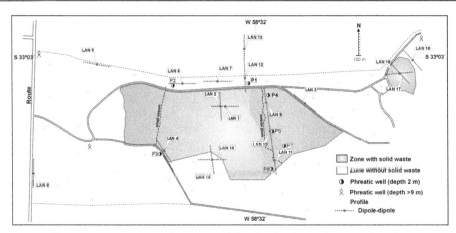

Fig. 10. A schematic map showing locations of dipole-dipole profiles (Lan) and phreatimeters (P).

2.3.2 Geological and hydrogeological setting

This region forms part of the Chaco-Paranense plain (Russo et al., 1979). The most important outcropping formations are described following Iriondo (1980) and Fili (2001) in Table 3.

AGE	UNIT	LITOLOGHY	THICKNESS (m)	HYDROGEOLOGIC BEHAVIOUR
Holocene	La Picada Formation	Silt and dark clay Fine quartzoses sandstones	1.5 to 3	Discontinuos unconfined aquifer, contains water table Poor water quality
Pleistocene	Punta Gorda Group	*Upper Section*: Brown Silt and clay with calcareous concretions and calcareous hardpan layers ("tosca") *Lower Section*: Clay with some gypsum crystals	20 to 40	Low productivity aquifer to aquitard Unconfined to semiconfined, contains water table Good to regular water quality
Upper Pliocene	Salto Chico Formation	Fine sandstones with silt and clay interlayers	60	Aquifer Excellent water quality

Table 3. Stratigraphic column.

2.3.3 Field work

Several geoelectrical studies, electrical tomography and vertical sounding using dipole-dipole and Schlumberger electrode configurations respectively and GPR profiles (antennas 150MHz -500MHz) were performed within and outside the landfill (Figure 9). Spatially coincident profiles for multi-electrode resistivity and GPR were carried out. A scheme of the distribution of the profiles in the studied area is shown in Figure 9. The profiles Lan01, Lan02 and Lan04 are located inside the landfill area, Lan14 and Lan15 are partially located inside the landfill, Lan09, Lan10 and Lan11 are closed the stream E where there are no waste deposits, while the profiles Lan03, Lan05, Lan06, Lan07, Lan08, lan12 and Lan13 are outside the landfill and around the waste deposits (see Figure10).

During 2005, seven small wells (phreatimeters) were dug to 2.30 m below surface outside the landfill (Figure 10). They were cased with PVC tubes, capped at both extremes and had gravel filters at the bottom. These phreatimeters are used to control water level, chemical composition and isotope content. These wells are located, two in the south border, two in the north border and the others are between the central sector and east sector (Figure 10).

Phreatic levels respond faster to precipitation. The direction of the phreatic system flow is N-NE toward the El Cura stream and the Gualeguaychú River, both of effluent character. Temperature and electrical conductivity were periodically measured *in situ*.

2.3.4 Results

2D resistivity models corresponding to Lan01, Lan07 and Lan15 profiles were selected to discuss the most prominent results. These have been plotted between 0 and 20 m depth. It was considered that this depth is most appropriate to describe the surface geoelectrical characteristics and also to compare all the models to the same depth. Important conductivity anomalies have been detected below the waste disposal. The 2D model obtained within the landfill (Lan01, Lan02 and Lan04) shows a shallow layer with a thickness of 2 m to 3 m and a resistivity range from 100 to 1000 ohm-m, this layer represents the waste deposit (Figure 11). The second layer has a thickness of 4 m to 5 m and a low value of resistivity of 3 ohm-m to 6 ohm-m, which is attributed to the contaminated zone. The third layer has a thickness of more than 10 m and a resistivity of 15 ohm-meter. Below this depth is observed another conductive layer (5 ohm-m).

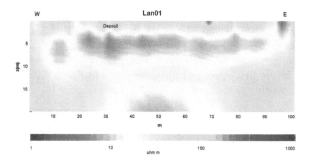

Fig. 11. 2D resistivity model up to 20 m deep obtained within the landfill (Lan01).

Several GPR profiles were performed inside and outside the landfill using a 150 MHz antenna and NS orientation covering approximately the entire length of the landfill and nearby outdoor areas.

The 2D electrical resistivity model and the GPR profile (150Mhz) of Lan02 are presented in Figures 12 and 13, and in this case the maximum depth of the electrical model was adapted to the result of GPR. The reflectors are at the depths where the waste is observed in the electrical model and then the signal is attenuated significantly. A conductive body is observed in the same place where there is a shadow zone in the GPR profile. The high conductivity of materials is attributed to fluids rich in salts, which attenuate the radar waves, not allowing them to reach great depths.

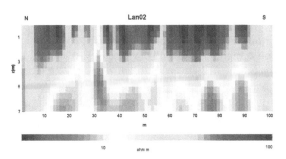

Fig. 12. 2D resistivity model up to 7m deep obtained within the landfill (Lan02).

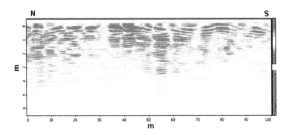

Fig. 13. Radar image for the landfill up to 7m of depth (Lan 02).

The 2D models obtained to the north of the landfill and outside its boundaries (Lan05, Lan06 and Lan07) also show zones with low resistivity values in the first 5 meters. This resistivity anomaly can be explained as leachate and also by the presence of clay. The resistivity bodies found at a depth around 10 m have been interpreted as calcareous concretions and calcareous hardpan layers that appear at different levels in Punta Gorda sediments (locally known as "tosca") according to the geological description (Figure 14).

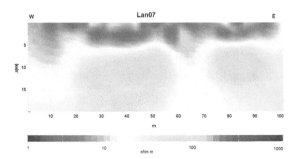

Fig. 14. 2D resistivity model up to 20 m deep obtained outside the landfill (Lan07).

Furthermore, the result corresponding to VES2 is presented in Figure 15. This sounding was performed close to Lan07 and shows similar results. It presents a conductive layer followed by another more resistive layer (15 ohm-m -25 ohm-m) extending up to near the 150 metres. The results are comparable because this model shows a first layer with a thickness of 2 m and resistivity of 4 ohm-m and a second layer with a thickness of 20 m and resistivity around 14 ohm-meters. Therefore, considering that the conductive layer is observed very superficially, the presence of leachate contamination is only restricted to the shallowest layer (<5 metres).

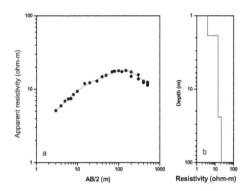

Fig. 15. a) Sounding curve, apparent resistivity vs. the half distance between the current electrodes. b) 1D model, depth vs. resistivity (log-log, representation).

Lan14 and Lan15 profiles are located in the southern part of the landfill. These profiles were made 30% over waste and the remaining 70% outside of the landfill. 2D electrical models show that in the surface and in the wastes the resistivity is high, about 250 ohm-m and a thickness around 3 meters. This result is consistent with Lan01, Lan02 and Lan04 models, which were discussed above showing that the resistivity is higher at the shallow depth where the wastes are deposited. Below them there is a conductive body with a resistive around 6 ohm-m which continues to the surface in the waste-free part. Where there is no waste the resistivity values varies between 3 ohm-m and 6 ohm-m with a thickness ranging from 2 m to 5 meters. Below this layer the resistivity ranges from 20 ohm-m to 40 ohm-meters (Figure 16).

Fig. 16. 2D resistivity model up to 20 m deep obtained outside the landfill (Lan15).

2.3.4.1 Electrical conductivity of phreatic and surface water

The water electric conductivity measured in the phreatimeters installed near the border outside the landfill was related with the electric conductivity measured in the geoelectrical models obtained from tomography.

The electrical conductivity of phreatic, surface water and effluents were measured. Values measured in effluents and surface water bodies (accumulations between cells, small streams, temporary ponds, etc.) range between 170 µS/cm and 800 µS/cm and the measures in phreatimeters between 240 µS/cm and 8300 µS/cm. An increase is noticed from south to north and from SE to NW, in the direction of the regional flow. These values are strongly influenced by precipitations and a dilution is observed after each rain event (Dapeña, pers. Com.). In addition, effluents of the landfill show a fast response to the influence of precipitations.

2.3.4.2 Calculations of hydrochemical parameters

Table 4 shows the hydrochemical parameters calculated for the oldest landfill. The depth of the minimum resistivity was determined in the central part of the model (x=50m) of profiles Lan01, Lan02 and Lan04. In the case of Lan14 and Lan15 the analysis was only made in the section of the profile which contains solid wastes and the depth of the minimum resistivity was established.

The procedures illustrate by Meju (2000) were used for the calculations and the hydrochemical parameters are shown in Table 4.

The results show the values calculated using equation (2) are slightly more than double those calculated using equation (1).

Although the results presented in Table 4 are approximate, these values can be considered as maximum and minimum values of prediction in this study and are useful for evaluating the current environmental conditions. It must be taken into account that these results are an approximation and also that the parameters depend upon the leachate composition, infiltration, waste types and geological materials.

The age of this landfill was inferred using the minimum resistivity values obtained in the models (Table 4). It is possible to define a range between 5 years and 20 years which matches with the age of the municipal landfill, since it was completely closed at least 5 years before this study was done.

The profiles were located close to the phreatimeters. Table 5 shows the comparison between resistivity measured in the phreatimeters and resistivity determined by the geophysical models at the same point and depth.

The expressions of Meju (2000) were also applied to estimate the electrical conductivity and resistivity of the fluid. The estimated resistivity is significantly lower than the experimental resistivity. This discrepancy can be explained by the presence of several levels of clay, which Meju's equations do not take into account (Table 5).

Additionally, groundwater electrical conductivity measured in the phreatimeters was compared to phreatic aquifer electrical conductivity obtained from the 2D model at the same point and depth. A good lineal correlation was observed between both conductivities. These indicate that the formation factor is almost homogeneous in the aquifer.

Profiles	σb (mS/m) ρb (ohm m)	TDS Predicted (mg/L)	σw Predicted (mS/m)	Cl- Predicted (mg/L)	W Estimate (Vol %)	Age (Year)	Profiles	σb (mS/m) ρb (ohm m)	TDS Predicted (mg/L)	σw Predicted (mS/m)	Cl- Predicted (mg/L)	W Estimat e (Vol %)	Age (Year)
Lan01 (a) x = 50M z = 4,5M	169,49 (5,9)	3944,58	568,43	1121,02	52,89	10-20	Lan 01 (b) x = 50M z = 4,5M	169,49 (5,9)	9342,64	1329,28	3106,98	33,82	5-10
Lan 02 (a) x = 50M z = 6M	109,29 (9,15)	2124,89	309,02	539,49	57,86	10-20	Lan 02 (b) x = 50M z = 6M	109,29 (9,15)	4733,25	680,25	1390,60	38,19	10-20
Lan 04 (a) x = 50M z = 5M	135,50 (7,38)	2877,00	416,53	771,92	55,37	10-20	Lan 04 (b) x = 50M z = 5M	135,50 (7,38)	6604,41	944,51	2061,80	35,98	5-10
Lan 14 (a) x = 74,5M z = 4,7M	197,90 (5,05)	4907,87	704,97	1451,46	51,25	10-20	Lan 14 (b) x = 74,5M z = 4,7M	197,90 (5,05)	11878,80	1684,13	4127,17	32,40	0-5
Lan 15 (a) x = 17,4M z = 6,7M	163,99 (6,10)	3765,27	542,97	1061,04	53,25	10-20	Lan 15 (b) x = 17,4M z = 6,7M	163,99 (6,10)	8876,89	1263,96	2924,70	34,13	5-10

Table 4. Hydrochemical parameters calculated for the landfill. The values of minimum electrical resistivity obtained ρ_b were used to predict TDS (a) with equation (1) and (b) with equation (2); the electrical conductivity of fluid is predicted using equation (3) and the chloride content with equation (4); The water content W is estimated applying equation (5) and the years are inferred using the content of TDS and Cl- (Meju, 2000).

Phreatimeter	Profile	Depth (m)	σwm (date) (mS/m), (Month/Year)	ρwm (ohm m)	ρb (ohm m)	ρw Predicted (ohm m) (a) (b)
P1	Lan07	2,15	204,0 (4/05)	4,90	5,75	0,72 - 3,22
P1	Lan12	2,15	136 (12/06)	7,35	9,12	1,46 - 3,22
P2	Lan06	2,10	174,3 (4/05)	5,73	6,90	0,96 - 2,19
P4	Lan09	1,76	267,0 (12/06)	3,74	5,43	0,66 - 1,57
P5	Lan10	1,45	490,0 (12/06)	2,04	3,14	0,29 - 0,73
P6	Lan11	2,00	75,2 (12/06)	13,29	11,63	2,12 - 4,52
P7	Lan11	2,05	123,6 (12/06)	8,09	8,27	1,26 - 2,81

Table 5. Comparison between resistivity measured in phreatimetres (ρ_{wm}) and resistivity determined in geophysical models (ρ_b) at the same point and depth. The last column shows water resistivity applying (a) equation 1 and (b) equation 2

2.3.5 Gualeguaychú landfill conclusions

The 2D models obtained within the landfill show a first layer with high resistivity (100 to 1000 ohm-m) and a thickness of less than 4 meters. This layer hosts the domestic and industrial wastes.

The low resistivity values observed below the wastes which could reach up to 10 m of depth are possibly due to leachate retained in the sediments of the Punta Gorda Group.

The neighbouring areas also have very conductive shallow zones which may be due to the migration of the leachate through groundwater flow or to the lithology of the deposits. At the northern boundary of the landfill a conductivity surface was detected and relatively high values of conductivity were measured in the wells. Some resistive bodies (~100 ohm-m) were found enclosed in a layer with a resistivity in the order of ~15 ohm-m. These anomalies are explained as lithological variations of the Punta Gorda Group.

The eastern boundary of the landfill has a conductive layer located at the same level as the layer inside. This could be due to the contamination at the border of the waste deposit and also by the presence of clays.

The results of GPR profiles showed strong reflectors, where the wastes are deposited and the upper limit of the contaminant plume was identified along the profile by the absence of reflectors or the existence of very weak signals. In this zone, the 2D resistivity model shows the presence of high electrical conductivity materials, which do not allow radar waves to reach greater depths.

The dipole-dipole and GPR results show good agreement and the integrated interpretation, is supported by local geology.

In the most conductive zones below the landfill some hydrochemical parameters of the leachate were predicted applying the empirical relationships defined by Farquhar (1989) and Meju (2000). The fluid conductivity values are in general comparable with the electrical conductivity measured in water surface samples and less than in the water samples from phreatimeters.

2.4 Bariloche Landfill

2.4.1 Background

The landfill of the city of San Carlos de Bariloche is located on National Route 258 (RN 258) approximately 8 km SW. The site was initially used as a place of extraction of materials for the construction of the road (Figure 17). The generated depression was used for disposal of wastes. It is an open dump around 10 hectares area and 8 m to 10 m deep, which began operations in the early 80 s.

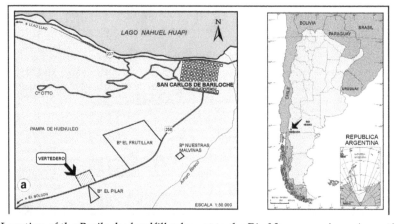

Fig. 17. Location of the Bariloche landfill relevant to the Rio Negro province, Argentina.

The deposition substrate is formed by glacial deposits, highly permeable and with low sorting. It is mainly composed of blocks, gravel and sand. Due to these textural characteristics, leachate has high potential to reach and contaminate the aquifers. Groundwater is the source of drinking water for urban areas adjacent to the repository.

2.4.2 Gelogical and hydrogeological setting

Outcrops present in the area correspond to geological units that comprise the Nahuel Huapi Group of Lower Tertiary age and the sedimentary Complex post-Pliocene-Quaternary which were descript by González Bonorino (1973) and González Bonorino & González Bonorino (1978) and are presented in Table 6.

AGE	UNIT		LITOLOGHY	HYDROGEOLOGIC BEHAVIOUR
LowerTertiary - Quaternary	Sedimentary Complex		Glacial and glacifluvial sediments	Aquifer, high to moderate permeability
Upper Tertiary	Nahuel Huapi Group	Ñirihuau Formation	Feldspar Wackes tuff, etc.	Some aquifer sections located in rocks with
		Ventana Formation	Lavas, brecchias, tobas, etc.	fissure porosity Very low permeability
Premesozoic	Cristalline Basament		Igneous and metamorphic rocks	Aquiclude

Table 6. Stratigraphic column

The landfill is located on sediments of the Sedimentary Complex post-Pliocene Quaternary, which also form a watershed and are part of the recharge zone of at least three major aquifer systems. These aquifer systems and related channels provide drinking water to urban areas nearby. Figure 18 shows a picture of the area.

Fig. 18. A photo of the landfill landscape

2.4.3 Field work

In the study area more than 30 GPR profiles were performed (Figure 19) reaching depths between 10 m and 40 m and 4 electrical tomographies using the DD configuration. DD1 and DD2 profiles were located within the landfill and DD3 and DD4 outside of it. Profiles DD1, DD2 and DD4 were oriented in the same direction (NW-SE), while the DD3 is almost perpendicular to them (see location in Figure 19).

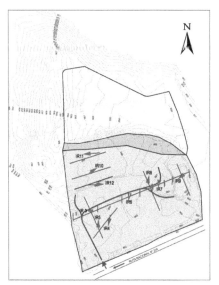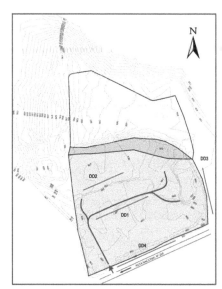

Fig. 19. A schematic map showing location of: a) GPR profiles; b) Geoelectrical profiles: DD1, DD2, DD3 y DD4.

2.4.4 Results and discussion

The lowest electrical resistivity obtained in the different models is about 1 ohm-m, while the highest is near the 10000 ohm-meters. These four orders of magnitude difference allowed the detection in the landfill areas of conductive anomalies which can be interpreted as groundwater contamination by leachate and the average resistivity of the ground outside of the landfill indicates the direction of flow of leachate.

Models of DD1 and DD2 profiles show variations in the electrical resistivity from 1 ohm-m to 1000 ohm-m (Figure 20 and 21). In both cases, the first 3 m - 4 m are very resistive. This layer represents the waste deposit and shows heterogeneity typical of areas with higher air content. These resistivity anomalies could be indicating the maximum depth of the deposit. Below the waste the resistivity values range from 20 ohm-m to 60 ohm-m reaching depths around 7 m to 12 m and the more conductive zones observed up to 20 m deep especially at the ends of the two profiles could be due to the presence of leachates. In the case of DD2 profile, where the anomaly is more extended, this could be explained by the uncontrolled discharging of waste fluids.

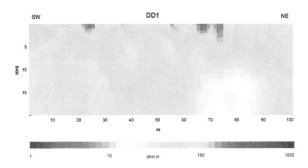

Fig. 20. 2D resistivity model up to 20 m deep obtained within the landfill (DD1).

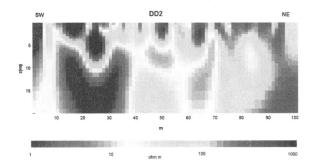

Fig. 21. 2D resistivity model up to 20 m deep obtained within the landfill (DD2).

Models of the DD3 and DD4 profiles show that the resistivity of the shallow layer for both cases is about 5000 ohm-m - 2000 ohm-m, but below 10 m deep the minimum resistivity is around 200 ohm-m for DD3 while in DD4 the resistivity anomaly is below 20 ohm-meters (see Figures 22 and 23).

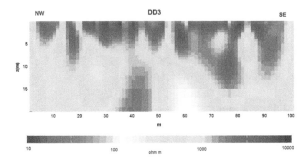

Fig. 22. 2D resistivity model up to 20 m deep obtained outside the landfill (DD3).

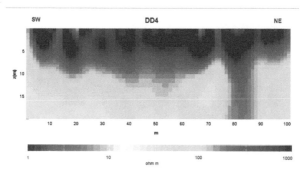

Fig. 23. 2D resistivity model up to 20 m deep obtained outside the landfill (DD4).

This behaviour suggests that the leachate could be migrating to this area (RN 258) and contaminating it corresponding to the DD4 profile. The zone corresponding to DD3 profile is uncontaminated. This is an important conclusion, because it coincides with the general idea of the flow direction provided by previous investigations.

Analysis of the GPR profiles has shown the characteristics of the fill, which is irregular in thickness, layout and compaction. Figure 24 presents the radar image of IR4 with 150 Mhz antenna showing the structure of the filling to a depth of 3 m and distinguishes the presence of an object buried at shallow depths.

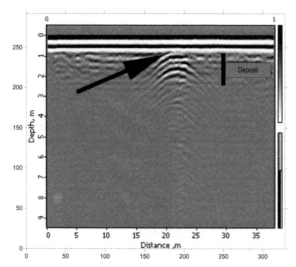

Fig. 24. Radar image for the landfill up to 9 m of depth.

2.4.5 Bariloche Landfill conclusions

In summary, this study reflects the existence of an irregular filling with variable depths between 5 m and 7 m, with a maximum depth estimated of 12 meters. The presence of leachate, recognizable to a depth of around 20 m - 25 m, suggests that levels are not impermeable so as to prevent the migration of leachate into the subsoil. The contamination plume moves towards to the RN 258, there is no recorded evidence of its displacement in NE direction.

Electrical resistivity models obtained are characteristic of landfills with large resistivity contrasts due to the presence of highly conductive fluids and waste which are usually resistive. These results have been correlated with images obtained with GPR. Also, the presence of buried objects of considerable size, associated with piece of metal, oil drums, amongst other things, has been detected.

3. Conclusion

In both cases studied, geoelectrical methods and ground penetrating radar have shown that they are efficient techniques for identifying contamination plumes produced by waste disposal sites due to their high salinity.

These non-invasive geophysical methods proved to be useful in terms of detecting and mapping physical-chemical changes associated with the presence of wastes. These methods do not require drilling and so avoid the risk of further contamination of the surface and groundwater.

Geological, hydrogeological and hydrochemical data are required to complement the geophysical results providing a more complete interpretation.

The application of this methodology is promising for use in environmental impact assessments of sanitary landfills which are the most common way to eliminate solid urban wastes.

4. Acknowledgements

We wish to thank Eduardo Llambías and Gabriel Giordanengo for their field technical assistance. This research was supported by the Agencia Nacional de Promoción Científica y Tecnológica (PICT 2002, 12243) and the article processing charge was paid by Technological Transfer Office CONICET.

5. References

Bell, F. G. & Jermy, C. A. (1995). A seepage problem associated with an old landfill in the greater Durban area. In: Sarsby, R.W. (Ed.). *Waste Disposal by Landfill GREEN'93*: 607–614. A.A. Balkema, Rotterdam.

Birks, J. & Eyles, C. A. (1997). Leachate from landfills along the Niagara Escarpment. En: Eyles, N. (Ed.). *Environmental Geology of Urban Areas. Geological. Associaiton of Canada*, Chap. 24: 347–363. Canada.

Buselli, G., Barber, C., Davis, G. B. & Salama, R. B. (1990). Detection of groundwater contamination near waste disposal sites with transient electromagnetic and electrical methods. En: Ward, S.H. (Ed.). *Geotechnical Environmental Geophysics*. Vol. 2: 27-39. SEG Publ, Tulsa, OK.

Bobachev, A., Modin, I. N. & Shevnin V. (1990-2000). Software. Department of Geophysics. Geological Faculty. Moscow State University y Geoscan-M. Ltd.

Daniels, D. (2004). Ground Penetrating Radar. 2nd Edition. *IEE Radar, Sonar and Navigation Series*, 15, 726 pp.

Farquhar, G. J. (1989). Leachate: production and characterisation. *Canadian Journal of Civil Engineering* 16: 317–325.

Fili, M. (2001). Síntesis geológica e hidrogeológica del noroeste de la provincia de Entre Ríos – República Argentina. *Boletín Geológico y Minero*, Vol 112, Número especial: 25-36. ISSN 0366-0176. Madrid.

Iriondo, M. M. (1980). El Cuaternario de Entre Ríos. *Ciencias Naturales del Litoral, Revista* N° 11: 125-141, Santa Fé. Argentina.

González Bonorino, F. (1973). Geología del área entre San Carlos de Bariloche y Llao Llao, provincia de Río Negro. Departamento Recursos Naturales y Energía, Fundación Bariloche, Publicación 16: 53p.

González Bonorino, F. & González Bonorino, G. (1978). Geología de la región de San Carlos de Bariloche: Un estudio de las Formaciones Terciarias del Grupo Nahuel Huapi. *Asociación Geológica Argentina, Revista* XXXIII (3): 175-210.

Loke, M.H. (1999). Time-lapse resistivity imaging inversion. *Proceedings of the 5th Meeting of the Environmental and Engineering Geophysical Society European Section, Em1.*

Meju, M. (2000). Geoelectrical investigation of old abandoned, covered landfill sites in urban areas: model development with a genetic diagnosis approach. *Journal of Applied Geophysics* 44: 115–150.

Oldenburg, D. W., McGillivary, P. R. & Ellis, R. G. (1993). Generalized subspace method for large scale inverse problems. *Geophysics* V.114, p.12-20. 1993.

Oldenburg, D. W. &Li, Y. (1994). Inversion of induced polarization data. *Geophysics* V59: 1327-1341.

Radar System, Inc. Software Prism. (2004). http://www.radsys.lv.

Pomposiello, C., Favetto, A., Boujon, P., Dapeña, C. & Ostera, A. (2008). Evaluación del estado actual de sitios de disposición final de residuos sólidos urbanos aplicando técnicas geofí-sicas. *Revista de Geología Aplicada a la Ingeniería y al Ambiente.* Vol 22 123-133. ISSN 1851-7838.

Pomposiello, C., Dapeña, C., Boujon, P. & Favetto, A. (2009). Tomografías eléctricas en el basurero municipal ciudad de Gualeguaychú, provincia de Entre Ríos, Argentina. Evidencias de contaminación. *Asociación Geológica Argentina, Revista* 64 (4) 603-614.

Prezzi, C. Orgeira, M. J., Vásquez, C. J. A & Ostera, H. (2005). Groung Magnetic of a municipal solid waste landfill: pilot study in Argentina. *Environ. Geol.* 47: 889-897.

Radar System, Inc., 2004. Software Prism. http://www.radsys.lv.

Russo, A., Ferello, R. E. & Chebli, G. (1979). Cuenca Chaco Pampeana. En: Geología Regional Argentina, II Simposio de Geología Regional Argentina. *Academia Nacional de Ciencias de Córdoba*, Vol. I (4): 139-183. Córdoba.

Yaramanci, U. (1994). Relation of in situ resistivity to water content in rock salts. *Geophysical Prospecting* V. 41, p. 229–239.

3

Integrated Study on the Distribution of Contamination Flow Path at a Waste Disposal Site in Malaysia

Kamarudin Samuding[1], Mohd Tadza Abdul Rahman[1],
Ismail Abustan[2], Lakam Mejus[1] and Roslanzairi Mostapa[1]
[1]Malaysian Nuclear Agency (Nuclear Malaysia),
Bangi, Kajang, Selangor,
[2]School of Civil Engineering, Universiti Sains Malaysia,
Nibong Tebal, Penang,
Malaysia

1. Introduction

Generally, the amount of solid waste generation is increasing as the economy and population continue to grow all around the world. The world's total solid waste generation was about 12.7 billion tonnes in 2000, and this is predicted to rise to about 19.0 billion tonnes in 2025 (Yoshizawa et al. 2004). In the case of Malaysia, it is estimated that 17,000 tonnes of solid waste is generated every day, and this will increase to more than 30,000 tonnes per day by 2020 consequent upon the increasing population and per capita waste generation (MHLG, 2003). Recently, the per capita generation of solid waste in Malaysia varies with an average from 0.8 to 1.0 kg/day depending on the economic status of an area (MHLG, 2003). Fauziah and Agamuthu (2006) estimated that the generation rate of solid waste may be increased by 3% per year due to the increase in population and the economic development in the country.

According to Mitsuo et al. (2008), solid waste accumulated in waste disposal sites or landfills can be decomposed by a combination of chemical, physical, and biological processes. Those decomposition processes occur as infiltrative water percolates through the solid waste in the landfill. As a result, various organic and inorganic compounds leach out from the landfill. The products of the complex combination of reactions are potentially transported further by the percolating leachate. Thus landfill leachate contains many constituents including potentially toxic substances, and its quality is heterogeneous. In this case the migration provokes environmental pollution especially in the local subsurface zone and hydrosphere. This phenomenon can be found around open-dump sites.

Most of the waste disposal site in Malaysia can be categorized as open dump sites which are usually without proper liner, treatment facilities and final capping. Until 2008, there are 180 landfills still in operation (Aziz, 2009). Most of these landfills are poorly managed and as a consequence leachate will easily migrated to the surrounding area through soils, subsurface geological strata and finally to the groundwater. The high annual rainfall in Malaysia with an average of 3000mm (Department of Irrigation and Drainage 2000, unpublished) also influenced the generation of leachate at these landfills. This situation will give some impact especially to the soil and groundwater contamination beneath a landfill site and poses a continuing risk to human health and the environment. Liquid contaminants can migrate through the soil matrix and leach into groundwater, while solid and semi-solid pollutants may be transported and dispersed through the subsurface (GETF, 1996).

The problem of groundwater contamination by Waste disposal site is steadily growing worse in Malaysia due to the way of managing municipal solid wastes (Mohd Tadza et al, 1999). Previous studies that were carried out at waste disposal sites in Malaysia (such as Gemencheh and Pulau Burung) indicate that the quality of groundwater decreased due to the leachate movement into the groundwater system (Mohd. Tadza et al., 1999 and Mohd Tadza et al., 2005). High concentration of heavy metals such as lead, copper, nickel, cadmium and zinc can cause serious water pollution and threaten the environment (Aziz et al., 2004a; Ngah et al., 2008). To solve these problems, the contaminants must be removed or treated from the leachate (Kadirvelu et al., 2001).

The waste disposal are well known to release large amounts of organic and inorganic contaminants via leachate. In humid and semi-humid regions, leachate is produced primarily in association with precipitation that infiltrates through the refuse in landfill. Continuation of leachate generation at the landfill site will normally result in the migration of leachate plume into the underlying groundwater zone and pollutes it. A variety of heavy metals are frequently found in landfill leachate including, iron, zinc, copper, cadmium, lead, nickel, chromium and mercury (Ozturk et al., 2003; Aziz et al., 2004). In several instances, heavy metal concentrations in leachate increased by time because they are non-biodegradable and they can be accumulated in living tissues and causing various diseases and disorders (Wan Ngah and Hanafiah, 2008).

Since the refuse has the potential to contaminate the ground water system, there is a need to study the degree of pollution in groundwater and to assess the distribution and flowpath of pollutant species and their impact on water quality. In this chapter, integrated study with various approaches was conducted in order to determine the seriousness of the distribution and flow path of the contaminant to the surrounding area in a selected waste disposal site at Taiping, Perak. Malaysia.

2. Description of the site

The Taiping landfill is located in the state of Perak at $4°$ 49'N, $100°$ 41'E, covering an area of 50 acres (Figure 1). Since starting its operation in 1995, roughly about 660,000 metric tons (about 200 metric tons daily) of domestic wastes had been dumped in the area. The topography in the vicinity of the landfill is generally flat and low lying with local elevations at the site ranging from a high of 3.3m above sea level to a low of 1.8m. The climate of the area is classified as typical of Peninsula Malaysia (equatorial) characterized by uniform temperature (daily mean minimum and maximum of $30°C$ and $34°C$ respectively) and high humidity (80% - 90%).

This area is one of the wettest areas in Malaysia because of high average annual rainfall (an average of 4000mm). Larut River and its tributary Batu Tegoh River border the landfill site on the south and east respectively. The North-South Highway is just west of the site while at the north of the site is another pond and oil palm estate. The site is in a rural area, and has sparse vegetation and poor fauna. Geologically, the site is located in an area where the formation is of the Quaternary period consisting of mainly recent alluvium. The soil investigation carried out by a consultant in 1993 at the site showed that the soil consists of silty sand with tracers of gravel over a layer of sandy silty clay.

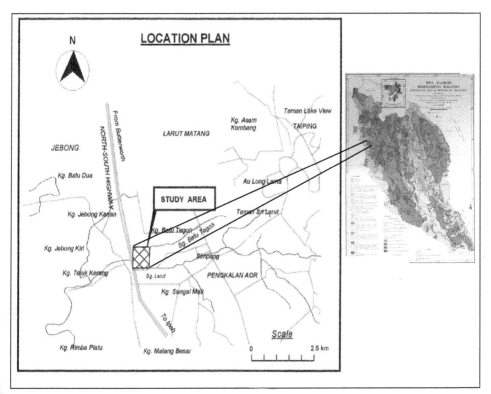

Fig. 1. Map of study area at Taiping waste disposal site.

3. Material and method

This chapter deals with field survey, sampling and laboratory test. Field survey involves geophysical investigation and groundwater flow study. In this study, electrical resistivity imaging (ERI) and colloidal boroscope system (CBS) were carried out to detect the flow path of leachate plume to the groundwater contamination at a waste disposal site. In addition, groundwater was sampled at every existing borehole within the study area in order to understand the scenario of the leachate plume distribution. The groundwater samples were analyse for their heavy metals content in the laboratory by using Inductively Couple Plasma Spectrometer (ICP-MS, model, Perkin – Elmer Optima 3000). Surfer software was used to plot the contours of heavy metals concentrations within the study area. These findings will help Local Authorities to take some immediate action to improve the existing landfill site for instances improving the leachate treatment facility and upgrading the infrastructure inside the landfill site.

3.1 Electrical resistivity imaging

Groundwater contamination investigation at the study site begin with minimally intrusive technique, called initial field screening technique. This technique is less expensive than the more intrusive techniques such as soil borings, test pits, and well monitoring. One of the principal categories of initial field screening techniques is shallow or surface geophysical survey, which include electrical resistivity imaging (ERI) technique. Knowing the depth of interest and data density necessitate, the configuration setting should be highly sensitive to ground conditions.

Shallow geophysical investigation can be considered as effective and reliable approach for characterize landfill sites. Recently developed geophysical hardware and software tools provide the opportunity to image the vertical structure of a landfill and its geologic setting. Electrical methods with multiple arrays have been widely used to detect spread of contamination, conductive media and groundwater contamination monitoring (Mota et al. 2004; Rosqvist et al. 2003; Buselli and Kanglin Lu 2001; Buselli et al.,1999). This methods also used to identify the limits and thickness of the dumpsite, delineated the base of a landfill and mapping the geometries of the host sediments (Cardarelli and Bernabini 1997; De Iaco et al. 2003; Gilles 2006).

ERI utilizes the injection of electrical current directly into the ground through current electrodes. The resulting voltage potential difference is measured between a pair of potential electrodes. The current and the potential electrodes are generally arranged in a linear pattern (**Figure 2**). The apparent resistivity is the bulk average resistivity of all soils and rock influencing the flow of current. It is calculated by dividing the measured potential difference by the input current, and multiplying by a geometric factor. The geometric correction is based on the arrangement of the current electrode or transmitter and the potential electrode or receiver in relation to each other.

The RES2DINV.EXE software is used to process the measured data involving inversion and to determine a 2-D resistivity model (Loke and Barker 1996). The Wenner-L and Wenner-S arrays were chosen due to it highly sensitivity to vertical-horizontal changes and the combination has a good vertical resolution to image the contaminated groundwater boundaries. The ERI survey at this site was carried out using ABEM Terrameter SAS4000 connected to LUND electrode selector 464 system (ES464) (ABEM 1998a, 1998b).

In practice, a line of multiple electrodes is deployed across the land surface. Electrodes are sequentially activated as either current or potential electrodes, and apparent resitivities are determined for numerous overlapping electrode configurations. The Wenner array was chosen in this study for several reasons. It is a robust array in the presence of measurement noise. It is well suited to resolving horizontal structures because it is more sensitive to vertical changes in resistivity than to horizontal changes in resistivity (Loke 2003).

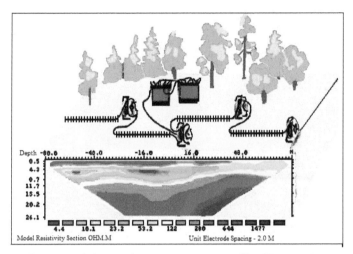

Fig. 2. The general setup and the resulting image processed by 2D inversion

3.2 Colloidal Borescope System (CBS)

The colloidal boroscope consists of two CCD (Charged-couple Device) cameras, a digital compass, an optical magnification lens, an illumination source and stainless steel housing. The device is approximately 89 cm long and has a diameter 44mm, thus facilitating insertion into a 50 mm diameter monitoring well. Data from the colloidal borescope is transferred to the camera control unit (CCU) at the surface by high strength electrical cable. The camera housing and light head are made of stainless steel, and are sealed for underwater used to 100 meter depths (**Figure 3**).

In this study, single well method was used to determine the groundwater velocity and flow direction. The colloidal borescope was inserted into the well at certain depth to monitor the movement of suspended particles. Upon insertion into a well, an electronic image magnified 140x was transmitted to the surface, where it was viewed by one of the CCD cameras in order to align the borescope in the well. As particles pass beneath the lens, the back lighting source illuminated the particles similar to a conventional microscope with lighted stage. A video frame grabber digitised individual video frames at intervals selected by the operator. AquaLITE Software package developed by Ridge National Laboratory compared the two digitized video frames, matched particles from the two images and assign pixel addresses to the particles. Using this information, the software programs computed and record the average particle size, number of particles, speed and direction.

When the colloidal borescope is inserted into a monitoring well, it directly measures the movement of colloids. With the insertion, the flow would initially swirl and manifest as multidirectional. If the borescope were moved after being placed into the well, swirling flow would continue. Consequently, it is necessary to secure the instrument cable on the surface to prevent movement of the borescope. Generally, after 20–30 min, laminar horizontal flow would dominate, and this could be observed in wells for certain periods of time. By plotting the trajectory and speed of colloidal particles across the screen with AquaLITE, the relative flow direction was determined .

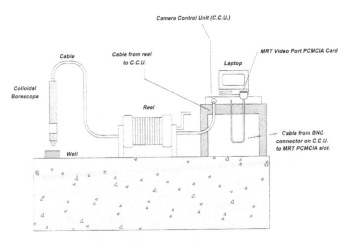

Fig. 3. Schematic diagram of Colloidal Borescope System

3.3 Hydro geochemical

Water sampling programme was conducted purposely to investigate the dispersion and flowpath of the pollutant species. A network of about twenty (20) observation points had been identified and collected for water samples that comprising of twelve (12) groundwater samples, three (3) river water samples, three (3) ex-mining pond and two (2) small streams.

Groundwater in all boreholes was sampled by using portable engine pump (Model Tanaka TCP 25B, maximum capacity: 110litres/min, maximum suction head: 8m and maximum delivery head: 40 m). Boreholes were pumped at least three well volumes before sampling to remove stagnant water in the borehole casing. Water samples for heavy metals analysis were collected in 1 liter High Density Polyethylene (HDPE) bottles which preserved with approximately 8 ml of 65% of nitric acid until the pH is < 2(Appelo and Postma , 1996). This process need to be follow in order to prevent the posibility of heavy metals cricipitated. The water samples were sent to a laboratory and analysed using the Perkin-Elmer Inductively Coupled Plasma Mass Spectrometry (ICP-MS) Model (Perkin Elmer Model ELAN 6000).

4. Results and discussion

In general this project had demonstrated the use of integrated techniques in assessing the distribution of contamination flow path at selected waste disposal site in Malaysia. This integrated study need to be conducted in order to get more conclusive results.

4.1 Electrical resistivity imaging

The electrodes spacing for ERI survey was set at 5m apart with the length of 200m, 300m and 400m. Such an arrangement would provide resistivity layer output of the subsurface geological information up to approximately 30m and 65m below ground surface respectively. In the study, a total of five 2-D resistivity survey lines were carried out in the survey (SL-1 – SL-5) in order to get data covering the dumping site and its surrounding area (**Figure 4**). For the comparison purposes, one survey line (SL 5) was conducted on the refuse itself (inside the waste disposal – contaminated area). Four survey lines (SL 1, SL2, SL 3 and

SL 4) were located at the outside the dumping site and one of the them was running parallel to the river at the south of the waste disposal site.

Figures 5 show the results of SL 5 which was laid inside the waste disposal show the possible occurrence of leacheate contamination near the surface down to 15 meters depth. This is due to the presence of low resistivity layer (< 10 ohm-m) blue in colour at the depth of about 5-15 meters. This finding is similar to Aaltonen and Olofsson's (2002) study showing that leacheate from the waste disposal has a low resistivity (about 1 ohm). The resistivity values (green colour 10-100 μm) normally indicate the existence of fresh water or sandy layer, while the highest resistivity values of (red color >100 Ohm.m) due to the backfill material as suggested by (Loke and Barker, 1996).

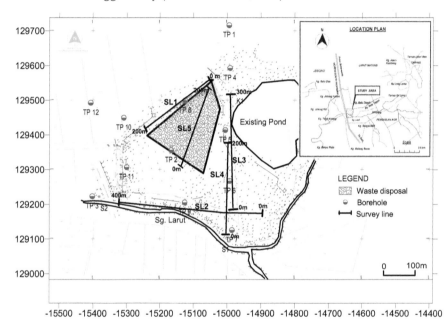

Fig. 4. Location of ERI survey line at the waste disposal site

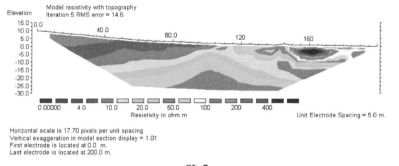

SL-5

Fig. 5. Profile of ERI survey line inside the waste disposal area

Similar low resistivity values of <10ohm-m can be seen more prominent distributed at SL3 and SL4 which were laid near the waste disposal site (**Figure 6**). This is indicated that the flow path of the leachate is moving towards to the southeast of the waste disposal site and can be infiltrated up to 30 meters depth. Meanwhile, the resistivity profile at SL1 is seen not much effected by the leachate plume. As stated earlier, SL2 was located outside the waste disposal boundary and parralel to the river. The results show that the resistivity values mostly indicate the existence of fresh water.

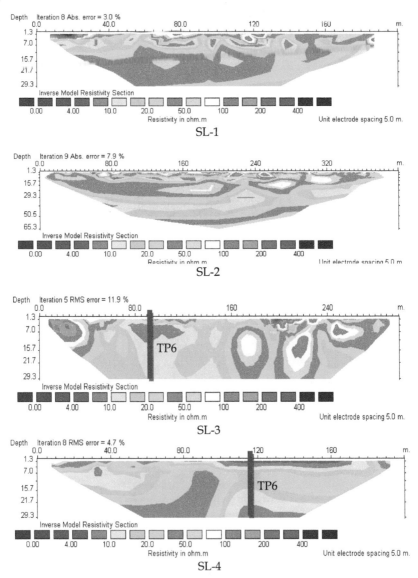

Fig. 6. Profile of ERI survey lines at the sorrounding of waste disposal area

As a summary, this study demonstrates that the electrical resistivity imaging (ERI) is viable tool for mapping groundwater contamination because electrical conductivity is directly related to the dissolved solute content in water. However, this data should be confirmed by groundwater movement and groundwater quality analysis within a particular hydrogeological strata by installing monitoring well.

4.2 Groundwater flow direction and velocity

The colloidal borescope system (CBS) was used at several boreholes at waste disposal site, Taiping, Perak. The purpose of this experiment was to determine groundwater flow pattern within the study area. Determining groundwater flow pattern was very important to obtain information on the migration and dispersion of pollutant materials seeping into the groundwater system. **Table 1** is a summary of the field results of groundwater flow velocity in the boreholes, as measured by the CBS. Based on results, the average of groundwater flow velocities is in the range of 1.09–3.86 x 10⁻⁴ m/sec. The various values of groundwater flow velocities were due to the difference of soil strata.

Figure 7 shows the regional and localized groundwater flow direction within the study area. Regional groundwater flow direction was obtained by using the conventional hydrological approach (i.e., by plotting the contour of groundwater table above mean sea level). From the plot, regional groundwater flow directions were quite scattered. For more detail, localized flow direction was obtained from colloidal borescope data. At the north of the study area the local groundwater flow is moving to the southeast and similar pattern can be seen at the center part of the study area. Whilst, at the south of the waste disposal site, which is bounded to the Sungai Larut, the local groundwater flow moved towards to the west and formed a localized groundwater, parallel to the river. Overall the localized flow directions dominantly flowed towards to southeast of the study site. These results can be correlated with the resistivity profile within the study site.

Boreholes	Velocity (m/sec)
TP1	3.86×10^{-4}
TP2	3.80×10^{-4}
TP3	2.74×10^{-4}
TP4	2.53×10^{-4}
TP5	1.09×10^{-4}
TP6	1.20×10^{-4}
TP7	1.89×10^{-4}
TP8	1.27×10^{-4}
TP9	1.28×10^{-4}
TP10	1.22×10^{-4}
TP11	1.09×10^{-4}
TP12	2.98×10^{-4}

Table 1. Data of groundwater flow velocity at the study area

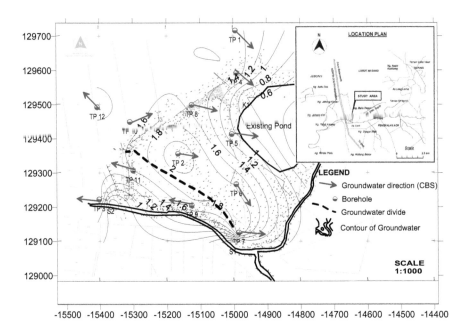

Fig. 7. Localize direction of groundwater movement within the study area

4.3 Flow path of pollutant species in groundwater

The average concentration of the pollutant species such as heavy metals in the groundwater system from several boreholes within the study area was obtained. A number of inorganic constituents detected in the examined samples indicated a small but significant presence of toxic materials. These data play an important role in the determination and visualization of the locations which are affected by the leachate plume. Hence, these results can help the local authorities to take action for remediation.

In this study, groundwater samples were analysed their heavy metals such as Pb, Cu, Fe and Cd. The concentration of pollutant species was plotted using the surfer software. **Figure 8-11** illustrate the flow path of pollutant species (i.e., Pb, Cu, Fe and Cd) in groundwater at the study area respectively. Based on the contouring diagram, the pollutants species seem to be accumulated within borehole TP6 that is located at the southeast of the waste disposal site. In other words, the pollutant species have a tendency to migrate and disperse toward the southeast of the waste disposal site, where the concentrations of pollutants species at this boreholes (TP6) is relatively high compared with other boreholes.

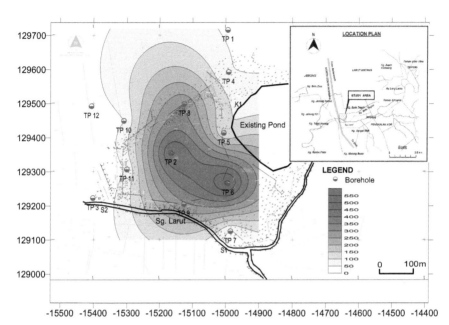

Fig. 8. Distribution of lead (Pb) at the study area

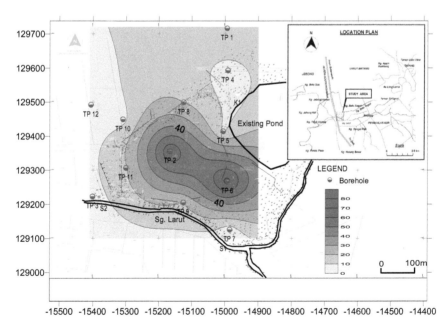

Fig. 9. Distribution of Copper (Cu) at the study area

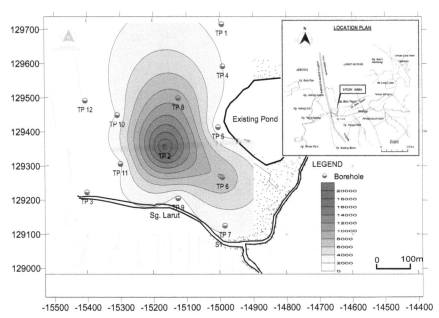

Fig. 10. Distribution of Iron (Fe) at the study area

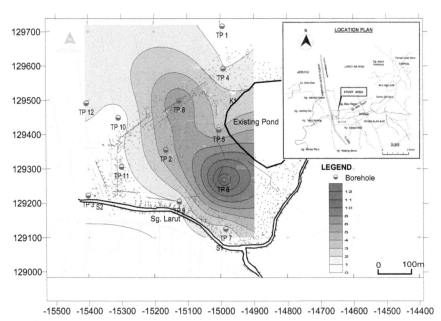

Fig. 11. Distribution of cadmium (Cd) at the study area

However, the distributions of contaminants were localized and confined within the dumping area and not diffuse over a large area. In addition, as previously mentioned that the groundwater flow direction measured by the colloidal borescope was dominantly towards southeast of the study area.

5. Conclusion

Leachate contamination at the Taiping waste disposal can be visually detected through ERI technique. In general, the contours of resistivity results show the existence of inhomogeneous strata in the area. It is quite clear that low resistivity anomalies exist at certain location in this study area is due to leachate plume movement. The result of the study confirms that the occurrence of groundwater contamination can be detected up to 30 m in-depth. The ERI technique had successfully delineated pollution layers. Thus, this method is an effective tool in detecting contaminated groundwater zones or layers in the study area.

With support from the colloidal borescope data, the movement direction of leachate plume can be determined. Generally, the flow pattern of the pollutant species dominantly towards to the southeast of the study area that is follow the flow direction of groundwater with flow velocity ranges between $1.09\text{-}3.86 \times 10^{-4}$ m/sec and it seems there is a possibility that the contaminant plume move slowly towards the Larut River.

Based on the geochemical analysis, higher anomaly pollutant species were detected at TP6 which is located at the southeast of the study area, indicates that the contaminant dominantly migrated through this borehole. However, the migration of leachate plume in the study area is still localized and not disperses in a wide area. This correlates well with low resistivity zone (<10 ohm-m) from the ERI images as shown in Figure 6.

Through this finding, it can assisst the Ministry of Housing and Local Government to formulate strategic and actions planning for improving the management and protection of water resources for long-term growth and sustainability. This can be done by;

i. Providing certain budget to Local Authorities to take some immediate action to improve the existing waste disposal site among others for instances improving the leachate treatment facility and upgrading the infrastructure inside the Waste disposal site. The improvement of the waste disposal at least up to level III of sanitary waste disposal system.

ii. Introducing solid waste management system that associated with the control of generation, storage, collection, transfer and transport, processing and finally disposing of solid wastes in a manner that is accordance with the best principles of public health, economics, engineering, conservation, aesthetics and environmental consideration.

6. Acknowledgment

First of all the authors acknowledge the Malaysian Government especially Ministry of Science, Technology and Innovation and Malaysian Nuclear Agency for their financial support. We would like to thanks to the staffs of Nuclear Malaysia especially to Hj. Juhari Yusof, Hj. Halim and others for their full commitment particularly in preparing and analysing the samples. We also extend our gratitude to Mr. Zainal Abidin Mat Yaman, Mr. Kamaruzzaman Kamari and Mr. Mohd Sharil Hj. Sharudin (Taiping Municipal Council) and others related to this study for their cooperation and technical assistance.

7. References

Aaltnen, J. and Olofsson B. 2002, 'Direct current (DC) resistivity measurements in long-term groundwater monitoring programmes', Environmental Geology, 41:662-671.

ABEM (1998a) Instruction manual, LUND Terrameter SAS 4000. Sweden: ABEM Instrument AB. Bromma. Atlas Copco.

ABEM (1998b) Instruction manual, LUND Imaging System. Sweden: ABEM Instrument AB. Bromma. Atlas Copco.

Appelo, C.A.J. and Postma, D., (1996). *Geochemistry, Groundwater and Pollution*. A.A Balkema, Rotterdam.

ARC Seibersdorf Research GmbH, (2003). User guide measurement system for determination of groundwater velocity and direction (Colloidal Borescope System).

Aziz, H.A., Yussuf, M.S., Adlan, M.N., Zahari M.S., & Alias, S., (2004a) Physico-chemical removal of iron from semi-aerobic landfill leachate by limestone filters. *Waste Management*. Vol. 24, pp 353-358.

Aziz, H.A., (2009) Semi aerobic Landfill, Penang Experience. *1st Regional Conference on Geo-Disaster Mitigation Waste Management in Asian*, Kuala Lumpur, 3-4 Mac 2009.

Benson RC, Turner M, Turner P, Vogel sang W 1988, 'In situ time-series measurements for long-term groundwater monitoring', In: Collins AG, Johnsons AI (ends) Ground-water contaminants: field methods, ASTM STP 963, Philadelphia, 58-77.

Busily G, Davis GB, Barber C, Height MI, Howard SHD 1992, 'The application of electromagnetic and electrical methods to groundwater problem in urban environments', Explore Geophysics 23, 543-555.

Buselli G, Lu KL (2001) Groundwater contamination monitoring with multichannel electrical and electromagnetic methods. Journal of Applied Geophysics 48: 11-23.

Colucci P and Lavagnalo MC. 1999, 'Three years of field experience in electrical control of synthetic waste disposal liners', Proceedings Sardinai '95, 5th International Waste disposal Symposium, 437-452.

Department of Irrigation and Drainage Malaysia (2000) *Urban Storm water Management Manual for Malaysia (Manual Saliran Mesra Alam)*, PNMB, Kuala Lumpur, Malaysia.

Engineering and Environmental Consultants, (1993). *Preliminary Environmental Impact Assessment for the Development of a Sanitary Waste disposal for Majlis Perbandaran Taiping*. Unpublished.

Fauziah S.H., Agamuthu P. (2005). "Pollution Impact of MSW Landfill Leachate." *Malaysian Journal of Science*. Vol. 24, pp 31-37.

(GETF) Global Environment and Technology Foundation. (1996) "Market Assessment Protective Underground Barrier Technologies." prepared for the U.S. Department of Energy.

Gilles Grand jean (2006) A seismic multi-approach method for characterizing contaminated sites. Journal of Applied Geophysics 58: 87–98.

Kardirvelu, K., Thamaraiselvi, K., Namasivayam, C., (2001) Removal of heavy metals from the industrial wastewater by adsorption onto activated carbon prepared from an agricultural solid waste. *Bioresour. Technol.* Vol. 76, pp 63-65.

Loke, M.H. and Barker, R.D., 1996. Rapid least-squares inversion of apparent resistivity pseudo sections by a quasi-Newton method. *Geophysical Prospecting*. 44: 131 – 152.

MHLG (2003) Solid Waste Management Report. Ministry of Housing and Local Government, Malaysia.

Mitsuo Yoshida, Hamadi Kallali and Ahmed Ghrabi, (2008) Subsurface contamination of potentially toxic elements caused by unlined landfill. Proceedings of APLAS Sapporo 2008. The 5th Asian-Pacific Landfill Symposium Sapporo, Hokkaido, Japan, October 22 – 24, 2008.

Mohd Tadza Abdul Rahman, Daud Mohamad, Abdul Rahim Samsudin and Tan Teong Hing (1999). Migration of pollutants in groundwater at a domestic waste disposal in Malaysia. A case study: Pollutants distribution and groundwater quality. Computer Aided Workshop on Groundwater Contamination. 18-26, November.

Mohd Tadza, A. R., Khairuddin A.R., Kamarudin S., Ismail A., and Ismail C. M., (2005) Application of isotope hydrology to determine the impact of leachate from Taiping municipal waste disposal site on groundwater quality in Malaysia. ITC/IAEA Final report.

Mota R, Mateus A, Marques FO, Goncalves MA, Figueiras J, Amaral H (2004) Granite fracturing and incipient pollution beneath a recent landfill facility as detected by geoelectrical surveys. Journal of Applied Geophysics 57: 11–22.

Ngah, W.W.S. and Hanafiah, M.A.K.M., (2008). Removal of heavy metal ions from wastewater by chemically modified plant wastes as adsorbents: A Review. *Journal of Bioresource Technology* 99(2008), pp 3935-3948.

Ozturk, N. and Bektas, T.E. (2004) Nitrate removal from aqueous solution by adsorption onto various materials. *Journal of Hazardous Materials*, B112, pp 155-162.

Rosqvist H, Dahlin T, Fourie A, Rohrs L, Bengtsson A, Larsson M (2003) Mapping of leachate plumes at two landfill sites in South Africa using geoelectrical imaging techniques. Proceedings Sardinia 2003. 9th International Waste Management and Landfill Symposium.

Yoshizawa, S., Tanaka,M. and Shekdar, A. (2004): Global Trends in Waste Generation, Global Symposium on Recycling, Waste Treatment and Clean Technology (REWAS 2004), Vol. 2, pp. 1541-1552.

Part 2

Nuclear Waste Disposal

4

A New Generation of Adsorbent Materials for Entrapping and Immobilizing Highly Mobile Radionuclides

Yifeng Wang, Huizhen Gao,
Andy Miller and Phillip Pohl
Sandia National Laboratories
USA

1. Introduction

The United States is now re-assessing its nuclear waste disposal policy and re-evaluating the option of moving away from the current once-through open fuel cycle to a closed fuel cycle. In a closed fuel cycle, used fuels will be reprocessed and useful components such as uranium or transuranics will be recovered for reuse (e.g., Bodansky, 2006). During this process, a variety of waste streams will be generated (NEA, 2006; Gombert, 2007). Immobilizing these waste streams into appropriate waste forms for either interim storage or long-term disposal is technically challenging (Peters and Ewing, 2007; Gombert, 2007). Highly volatile or soluble radionuclides such as iodine (^{129}I) and technetium (^{99}Tc) are particularly problematic, because both have long half-lives and can exist as gaseous or anionic species that are highly soluble and poorly sorbed by natural materials (Wang et al., 2003; Wang and Gao, 2006; Wang et al., 2007). Waste forms are probably the only engineered barrier to limit their release into a human-accessible environment after disposal. In addition, during the fuel reprocessing, a major fraction of volatile radionuclides will enter the gas phase and must be captured in the off-gas treatment. It is thus highly desirable to develop a material that can effectively capture these radionuclides and then be converted into a durable waste form.

In fuel reprocessing, spent fuels are first subjected to voloxidation and acid dissolution, during which 94% to 99% of iodine-129, together with other volatile radionuclides (^{14}C, ^{3}H, Kr and Xe), are released to the off-gas stream (Gombert, 2007). Various methods have been proposed for the recovery of volatile ^{129}I, including scrubbing with caustic or acid solutions and chemisorption on impregnated adsorbents (Rovnyi et al., 2002; Kato et al., 2002; Gombert, 2007). The proposed adsorbents include both natural and artificial porous materials like zeolite (e.g., mordenite), alumina, and silica gels loaded with metals (e.g., Ag, Cd, Pb) that form low-solubility iodides or iodates. However, several issues have been recognized regarding the effectiveness of these materials (Gombert, 2007). For example, based on adsorption models, iodine diffusion inside the zeolite particles was found to be slow, only about 2×10^{-14} cm^2/s, and likely to limit the adsorption process (Jubin, 1994; Gombert, 2007). The stability of silver-exchanged zeolite is also a concern. For example, silver-exchanged faujasite (AgX) decomposes in the presence of NO_x and water vapor. AgX also does not exhibit satisfactory thermal stability during regeneration. In addition, the fibrous nature of mordenite may present an inhalation hazard (Stephenson et al., 1999). Silver-impregnated alumina and silica capture iodine by forming AgI along grain boundaries. The high iodine leaching rate of these materials can be a potential issue if they are directly used for disposal. Currently, no path forward has been established for the disposition of these adsorbent materials. In addition, metal-organic frameworks (MOFs) have been found to have high sorption capacities and are tunable to different volatile elements (Sudik et al., 2006; Lee et al., 2007). However, the long term fate of the organic structure in a high heat/radiation field is unknown.

In this chapter, we present a new concept for the development of a next generation of high-performance radionuclide adsorbent materials for nuclear waste reprocessing and disposal. Based on this new concept, we have developed a suite of inorganic nanocomposite materials (SNL-NCP) that can effectively entrap various radionuclides including gaseous [129]I and anionic [99]Tc. Importantly, after the sorption of radionuclides, these materials can be easily converted into nanostructured waste forms, which are expected to have unprecedented flexibility to accommodate a wide range of radionuclides with high waste loadings and low leaching rates.

2. Nano-scale radionuclide immobilization and encapsulation

The development of SNL-NCP materials are based on the concept of nano-scale radionuclide immobilization and encapsulation (Fig. 1). For example, for capture of iodine, a nanoporous material that has a high sorption capability for iodine gas is first synthesized. The sorption capability can be further improved, as needed, through pore surface modification. The material is then used as an adsorbent to entrap gaseous iodine. The entrapped iodine is converted to a less volatile compound, for example, by reacting with silver to form AgI precipitates inside nanopores of the material. The idoine-loaded material is finally mixed with glass-forming constituents and calcined, resulting in a waste form in which I-bearing nanoparticles are embedded in a glass or crystalline matrix. More recent development of this technology allows us to avoid using Ag or other heavy metals for iodine immobilization and subsequent vitrification (see Section 5 of this chapter).

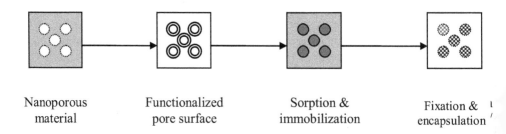

| Nanoporous material | Functionalized pore surface | Sorption & immobilization | Fixation & encapsulation |

Fig. 1. Formation of a nanocomposite waste form. In the final waste form, nano-scale radionuclide precipitates are embedded in a glass or ceramic matrix.

Existing waste forms tend to incorporate radionuclides into rigid crystal structures (e.g., Xu and Wang, 2000; Fortner et al., 2002; Ewing, 2006; Grambow, 2006), and the waste loading of those materials is generally limited by various structural factors such as the size and charge of a target radionuclide. Unlike the existing waste forms, the new waste forms we developed incorporate radionuclides as nano-scale inclusions in a host matrix and thus effectively relax the constraint of crystal structure on waste loadings. Therefore, the new waste forms are able to accommodate a wide spectrum of radionuclides with high waste loadings and low leaching rates. Since the leaching rates are controlled by the dissolution rate of the host mineral, it is even possible to engineer a waste form that will be thermo-dynamically stable in a repository environment, by choosing an appropriate host mineral. A good analog to this is the naturally occurring nano-scale fluid inclusions in mineral crystals, which remain intact over millions of years (Wang et., 2003). It is anticipated that SNL-NCP materials can be easily integrated into the fuel reprocessing system currently proposed for advanced fuel cycles.

3. Synthesis of nanoporous metal oxides

More than 300 nanoporous metal oxide materials have been synthesized with a sol-gel route developed specifically for this work. In this route, inorganic metal salts are used as precursors, a block copolymer as a nanostructural template, and an epoxy (propylene oxide in this case) as a proton scavenger. The synthesis can be conducted under relatively simple ambient conditions from room temperature to 65 °C. A nanoporous metal oxide is obtained by drying the resultant gel at 80 °C and calcining it at or below 600 °C in air or oxygen atmosphere. Since it uses inexpensive inorganic precursors, the synthesis route we developed can be scaled up for a large quantity of production if needed.

For example nanoporous alumina can be synthesized with the following chemicals and procedure. The chemcials used for the this synthesis include: aluminum chloride hexahydrate ($AlCl_3 \cdot 6H_2O$, e.g., from Fisher), propylene oxide (C_3H_6O, PO, e.g., from Fisher), poly(ethylene glycol)-block-poly(propylene glycol)-block-poly(ethylene glycol) (P123, e.g., from Aldrich, M_n=5800), and anhydrous alcohol (e.g., from Fisher Acros). For solution A, 12 grams of P123 are dissolved in 72 mL alcohol (anhydrous). For solution B, 34 g of $AlCl_3 \cdot 6H_2O$ is dissolved in 90 mL of a H_2O/C_2H_5OH (1:1) mixture. A typical synthesis process involves mixing the two prepared homogeneous solutions (A and B) until the solids dissolve, then combining them into a single parent solution. To this parent solution, 47 g of propylene oxide is added. This precipitates the alumina which is then aged at room temperature in the hood for about 3 days, and then dried at about 80°C in the oven for about 24 hours to turn it into a gel. The gel is finally calcined at 600°C for about 4 hours at a temperature ramp of about 5°C/ min.

The involvement of a structural directing agent and the control of the polymerization rate of alumina hydroxides are two important factors for this synthesis. Block copolymer P123 (or other copolymer) is used as a template for nanostructures. The polymerization rate is controlled by adding a certain amount of propylene oxide (or ethylene oxide or other epoxy) to solution B. Aluminum ion, Al^{3+}, in solution has a great tendency to form positively charged polynuclear species (hydroxoaluminum complex) as follows:

$$Al^{3+} + 2H_2O \longrightarrow Al(OH)^{2+} + H_3O^+ \ (\log K_1 = -5) \tag{1}$$

$$7Al^{3+} + 34H_2O \longrightarrow Al_7(OH)_{17}{}^{4+} + 17H_3O^+ \ (\log K = -48.8) \tag{2}$$

$$13Al^{3+} + 68H_2O \longrightarrow Al_{13}(OH)_{34}{}^{5+} + 34H_3O^+ \ (\log K = -97.4) \tag{3}$$

Propylene oxide in the solution acts as a proton scavenger by the nucleophilic reaction towards ring-opening. This reaction is irreversible and slow, thus maintaining a uniform pH gradient in the solution for hydrolyzing the aluminum salt.

The synthesis route has also been adapted to one-pot syntheses of nanoporous materials containing multiple metal components (e.g., Ag-Al, Ni-Al, or Bi-Al oxides). This one-pot process allows us to ensure the chemical and structural homogeneity of synthesized composite material. A nanoporous Ni-Al oxide material obtained using this route is shown in Figure 2. The structural homogeneity of this material is confirmed with transmission electron microscopic (TEM) observations.

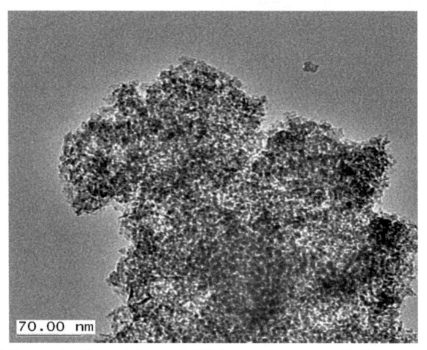

Fig. 2. Transmission Electron Microscopic (TEM) image of nanoporous Ni-Al oxides, showing worm-like nanopore structures.

The synthesized materials have been characterized with a powder x-ray diffractometer (XRD, Burker D8 Advance), surface area and porosity analyzer (BET TriStar 3000, Micromeritics), and transmission electron microscope (TEM, Jeol). The XRD analyses generally indicate that the synthesized materials have low crystallinity. The BET measurements show that these materials have pore sizes ranging from 11 to 16 nanometers and surface areas from 320 to 450 m^2/g (see Table 1, Fig. 3).

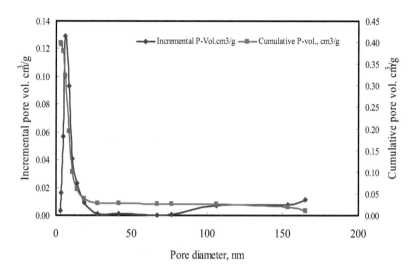

Fig. 3. Pore size distribution of Ni-Al (Ni:Al = 1:1) nanoporous material determined from BET measurements.

Sample ID	Single point surface area, m²/g	Pore volume, cm³/g Single point (ads.)	Pore size, nm (Ads. Average)
Nanoporous-alumina (commercial)	263.63	0.49	7.21
NC 13-A-600C/air	415.83	1.53	14.69
NC 13-A400C/O2	447.25	1.35	11.87
NC 13-B-600C/air	336.90	1.38	15.78
NC20-1 (m-Al batch 4)	319.35	1.13	13.65

Table 1. BET surface area, pore volume, and pore size of synthesized nanoporous alumina

4. Radionuclide sorption on nanoporous metal oxides

A set of SNL-NCP and other related materials were selected and tested for the sorption of gaseous iodine under simulated fuel reprocessing conditions. These tests provide important information regarding the selectivity of materials for iodine sorption and the potential interference of other chemical components anticipated to be present in off-gas waste stream from fuel reprocessing.

The tests were performed under elevated temperature (90°C) and variable relative humidity conditions, as well as with or without CO_2 and NO_2 gas added in the headspace. The general experimental setup for iodine sorption testing is shown in Figure 4. The Tc sorption capabilities were determined with batch experiments using Re as a chemical analog for Tc. The materials tested for iodine sorption include various SNL-NCP materials (NC-77), Al-Mg layered double hydroxides (NC-88) calcined at different temperatures, mesoporous silica, activated alumina (particles), sepiolite, palygorskite, and zeolite. These materials, all inorganic, encompass diverse chemical compositions and mineral structures and thus help to mechanistically understand iodine sorption on SNL NCP materials. After each iodine sorption test, the sorbent was mixed with SiO_2 and iodine concentrations in the adsorbent were measured with an ARL QUANT'X Energy-Dispersive X-Ray Fluorescence (EDXRF) Spectrometer (Thermo Electron Corporation).

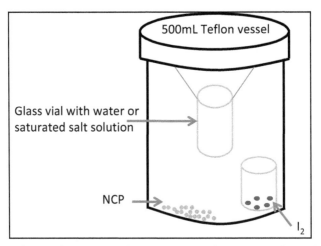

Fig. 4. Experimental setup for testing iodine sorption under various relative humidity conditions. The relative humidity inside the glass vial is controlled by a saturated salt solution.

Selected results of the batch sorption experiments are shown in Table 2. All these materials have exhibited high sorption capabilities for I_2 gas and TcO_4^- ion. A typical iodine sorption isotherm is shown in Figure 5. The data were obtained from exhaustion experiments, in which the iodine-loaded material was heated overnight at about 90 °C and cooled to about room temperature in an open jar. The change in the slope of the isotherm may reflect the transition from monolayer sorption to multi-layer sorption and eventually to pore condensation as the mass ratio of iodine to adsorbent increases. It was found that I sorbed in the Ni-Al material forms a separate Ni-Al-I phase in the nanopores, which may be responsible for the high I sorption capability observed. This leads to the possibility of using Ni or other metals, rather than relatively expensive Ag, for immobilizing I in a final waste form.

Materials	Specific surface area (m^2/g)	Pore size (nm)	Target radionuclides	Sorption capability
Alumina (Al_2O_3)	300-450	12-14	I	$[I] = 2 - 4x10^3$ ppm
Ag-Al oxide	200	6.5	I	$[I] = 1.9-4.0x10^4$ ppm
Ni-Al oxide	260	7	I, Tc*	$[I]=2x10^4$ ppm $K_d=210$ mL/g for Re
Bi-Al oxide	160	7	I, Tc*	$K_d=1680$ mL/g for Re

*Rhenium is used as the surrogate of Tc.

Table 2. Sorption capability of nanoporous metal oxides for iodine and technetium

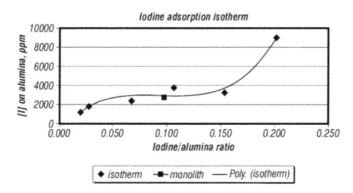

Fig. 5. Isotherm of iodine adsorption on nanoporous alumina. Measurements were made after overnight desorption at 90°C following adsorption

The experimental testing has confirmed that nanoporous Al oxide and its derivatives have high sorption capabilities for iodine sorption. To understand the underlying mechanism, the same sorption experiment was performed on nanoporous silica, which has a larger surface area and a smaller pore sizes than the alumina materials. Interestingly, no significant I sorption was observed on the silica material, implying that the surface chemical identity of the material plays an important role in I sorption. On the other hand, we have found that nanoporous structures can greatly enhance I sorption onto alumina. These observations lead us to conclude that the high I sorption capabilities of nanoporous alumina and its derivatives are attributed to the combined effects of surface chemistry and nanopore confinement. This points to a possibility of optimizing material performance for sorption capability and selectivity by manipulating material compositions and structures.

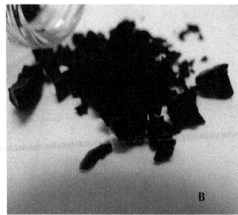

Fig. 6. Formation of monolithic nanoporous SNL-NCP-A. A – before I sorption; B – after I sorption. Each individual grain has nanoporous structures. The dark brown color in B indicates that the monolithic material has a high I sorption capability. The size of monolithic grains ranges from mm to cm.

SNL-NCP materials can also be engineered into monolithic forms (Fig. 6). Monolithic materials are generally preferred for engineering handling. As shown in Table 3, a monolithic form of SNL-NCP has an iodine sorption capability as high as the corresponding powder, indicating molecular diffusion in nanopores is not a limiting step for adsorption. Also, note that monolithic nanoporous alumina without silver incorporated has at least the same comparable iodine sorption capability as the nanoporous alumina with Ag included. Therefore, Ag may not be needed for I sequestration if nanoporous alumina is used as a getter material. Our experiments also show that zeolite materials have a low sorption capability for iodine, indicating that that zeolite component itself does not contribute much to the iodine sorption capability of Ag-exchanged zeolite materials and thus casting doubt on the necessity of using zeolite as a supporting material for iodine capture.

Material	I/(m-Al) ratio	Sample wt, g	[I] uptake, ppm	
Nanoporus alumina w/ Ag	0.114	0.2036	35674	
Monolithic Nanoporous alumina w/o Ag	0.107	0.2035	66245	
BET measurements				
Material	Surface area, m²/g	Pore vol. cm³/g	Pore size, nm	Micropore vol. cm³/g
Nanoporus alumina w/ Ag	215	0.706	12.7	0.006644
Monolithic Nanoporous alumina w/o Ag	354	1.75	19.15	0.014549

Table 3. Iodine sorption on nanoporous alumina and its derivatives

The experimental setup for iodine sorption tests under a controlled relative humidity is shown in Figure 4. The relative humidity (RH) in the headspace was controlled by the presence of a saturated salt solution in the test vessel. The salt solutions used include: LiCl - RH 10.23%, $MgCl_2$ - 24.12%, KCl - 78.5%, and deionized water (DI) - 100%. The testing results are shown in Figure 7. Nanoporous materials NC-77 and NC77-N2 outperform most of the other materials tested over the whole relative humidity range. SNL-NCP materials are relatively insensitive to water interference for iodine sorption. This unique property is important for the potential use of SNL-NCP materials as an iodine scavenger in the off-gas treatment during fuel reprocessing. Two layered double hydroxide materials NC88-380 and NC88-600, which can be considered as nanostructured layered materials, also exhibit high iodine sorption capabilities under high relative humidity. Sepiolite and palygorskite are two naturally occurring silicate materials with similar tunnel structures and slightly different chemical compositions. The two materials, however, display distinct sorption behaviors. Particularly, sepiolite is very sensitive to water vapor, even though its sorption capability for iodine is high at zero relative humidity. This is because sepiolite generally has a high affinity for polar molecules such as H_2O. Mesoporous silica has a relatively lower iodine sorption capability as compared with SNL-NCP materials with a similar nanoporous structure, again indicating the surface chemical identity of the material plays an important role in iodine sorption. It is interesting to note that zeolite has a relatively low sorption capability for iodine over the whole range of relative humidity.

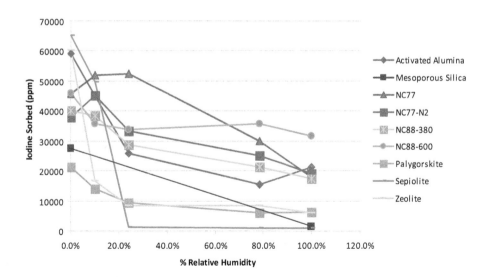

Fig. 7. Iodine sorption onto SNL-NCP and other related materials under variable relative humidity

5. Radionuclide fixation and encapsulation

A suite of techniques have been developed for the fixation of radionuclides in metal oxide nanopores. The key to this fixation is to chemically convert a target radionuclide into a less volatile or soluble form. Iodine can be fixed in two ways: (1) introducing Ag to a getter material to convert I_2 into AgI upon sorption and (2) adding a chemical base (e.g., Na_4SiO_4) to an I_2-loaded material to convert I_2 to iodide or iodate ions. X-ray photoelectron spectra (XPS) show that 66% of I_2 is converted to I^- upon iodine sorption onto Ag-alumina nanoporous material (Fig. 8). Similarly, Fourier Transformation Infrared spectroscopic (FTIR) analysis indicates that adding Na_4SiO_4 during I fixation forces the sorbed iodine gas in the nanopores to completely convert to I^- and IO_3^-, thus significantly reducing iodine volatility. The second approach is preferred for I fixation, because it avoids using silver. We have found that the same method may also apply to Tc fixation. Preliminary data indicate that addition of Na_4SiO_4 or NaOH helps convert TcO_4^- into insoluble TcO_2. The detailed mechanism for this conversion still needs to be clarified.

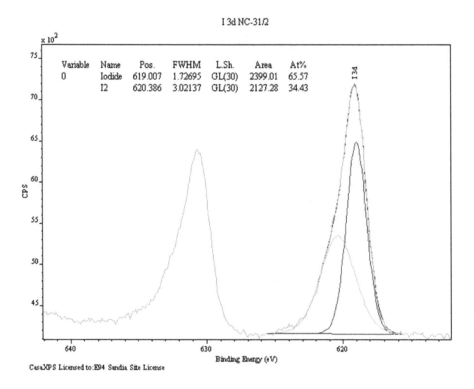

Fig. 8. X-ray photoelectron spectra (XPS) showing 66% of I_2 converted to I^- upon iodine sorption onto Ag-alumina nanoporous material.

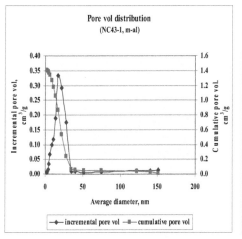
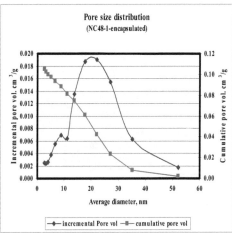

Fig. 9. Effective sealing of nanopores with Na-silicate solution. Left – pore volume distribution before sealing; Right - pore volume distribution after sealing. Note the changing y-axis values.

Radionuclides loaded onto a getter material are encapsulated by pore sealing and vitrification. We have found that mixing a radionuclide-loaded getter material with a certain amount of Na-silicate solution can effectively (>90%) seal the nanopores in the material (Fig. 9), thus enhancing radionuclide retention during subsequent vitirification of the material. No significant I loss was observed during vitrification if nanoporous alumina is used as an iodine getter.

A pore-sealed material is finally mixed with a glass-forming frit and vitrified to form a glass-ceramic waste form. We have tested six commercially available frits for their ability to isolate radionuclides. We have found that Fero frit "510" – a lithium borosilicate material – produces the best result in terms of waste form durability. Figure 10 shows some of glass-ceramic waste forms produced using the proposed nano-immobilization and nano-encapsulation technique. In these waste forms, radionuclide nanoparticles are embedded in either amorphous or crystalline matrix (Fig. 9).

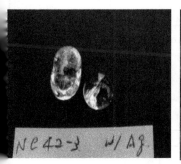
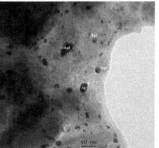

Fig. 10. Glass-ceramic waste forms produced using the proposed nano-scale encapsulation technique. Nanoparticles embedded: left – AgI-embedded glass –ceramic waste form; middle – TEM image of AgI-waste form; right – image of Bi-Tc waste form. Re is used as analog to Tc.

Furthermore, as shown in Table 4, with the fixation and encapsulation methods developed for this work, the loss of iodine during the conversion of an iodine-loaded nanoporous material into a waste form is minimal even at 1100 °C. It is clear that nanoporous structures not only enhance I sorption but also help to retain I during calcination (Table 5).

Glass sample	[I] in the ceramic-glass, ppm	[I], normalized to mass (g) of mesoporous alumina, ppm	vitrification temperature, °C	Iodine loss % during vitrification*
NC48-1+ "510"	429	7064	1100	32
NC48-2 + "510"	915	15067	1100	0
NC48-1 +"XF140-2"	698	11494	1100	0
NC48-2 +"XF140-2"	1069	17603	1100	0
NC52-2 + "3225"	617	9872	1100	5
NC52-2 + "CS749"	570	9120	1100	12
NC67-7	748	11968	1200	0
NC67-6	243	3880	1100	63
iso-750	855	13680	750	0
iso-800	649	10384	800	0
iso-850	659	10544	850	0
iso-900	706	11296	900	0
* Non-zero numbers are due to the heterogeneity of samples.				

Table 4. Iodine loadings on glass-ceramic materials synthesized with various glass forming frits under different process conditions

Material	I sorption (ppm)	% of I lost during fixation	% of I lost during vitrification
Alumina particles	98	Not tested	Not tested
Activated alumina particles	8700	45	65
Nanoporous alumina	25000	0	0

Table 5. Effects of nanoporous structures on I sorption and retention by alumina materials

6. Leaching tests of nanocomposite waste forms

The obtained nanocomposite waste forms were subjected to short-term leaching tests in deionized water. The results are summarized in Table 6. Three observations can be made from the table. First, nanoporous alumina (m-Al-I) fixed with potassium silicate provides the best performance in the leaching tests. Second, a material with no Ag incorporated performs better than the same materials fixed with silver. Therefore, silver may not be necessary for iodine immobilization in a glass-ceramic waste form. Third, there exists an optimal temperature for vitrification. This temperature is about 850-950 °C, which is lower than that used for glass formation (generally > 1100 °C). It is likely that the formation of nano-scale crystalline mineral phases at the optimal temperature may enhance waste form durability, which seems consistent with the high temperature X-ray diffraction (HTXRD) analysis (Fig. 11).

Material	pH-end	Iodine loss, %	vitrification T, °C	[SiO₂], ppm	composition
First leaching test (LA)					
LA-1	9.29	14.5	1100	not analyzed	m-Al-I+Na4SiO4+"510"
LA-2	9.50	27.4	1100	not analyzed	m-Al-Ag-I+Na4SiO4+"510"
LA-3	8.18	38.5	1100	not analyzed	m-Al-I+Na4SiO4+"XF140-2"
LA-4	8.22	33.3	1100	not analyzed	m-Al-Ag-I+Na4SiO4+"XF140-2"
m-Al-I /silver composite					
LB-1	8.40	37.4	1100	44	3225+NC52-2(vit),
LB-2	8.39	19.6	1100	44	CS749+NC52-2(vit)
LB-3	10.31	40.6	1100	717	m-Al-I + Na4SiO4
LB-4	10.66	29.6	1100	664	m-Al-I + Na4SiO4
LB-5	8.02	40.6	1100		m-Al-I+Na4SiO4 +SiO2+B2O3
LB-6	8.28	15.9	1200	27	m-Al-I+Na4SiO4 +SiO2
m-Al-I samples w/o silver					
LC-1	10.30	5.8	750		750 C. m-Al-I+Na4SiO4+"510" frit
LC2	10.12			1666	750 C. m-Al-I+Na4SiO4+"510" frit
LC-3	9.86	7.8	800	1034	800C. m-Al-I+Na4SiO4+"510" frit
LC-4	9.89			1013	800C. m-Al-I+Na4SiO4+"510" frit
LC-5	9.49	20.9	850	278	850C. m-Al-I+Na4SiO4+"510" frit
LC-6	9.52			419	850C. m-Al-I+Na4SiO4+"510" frit
LC-7	9.17	29.4	900	250	900C.m-Al-I+Na4SiO4+"510" frit
LC-8	9.18			213	900C. m-Al-I+Na4SiO4+"510" frit
m-Al-I samples fixed with potassium silicate					
LD-1	9.96	0.0	750	1444	750 C. m-Al-I+Na4SiO4+"510" frit
LD-2	10.02	0.0		2145	750 C. m-Al-I+Na4SiO4+"510" frit
LD-3	9.74	0.0	800	987	800C. m-Al-I+Na4SiO4+"510" frit
LD-4	9.52	0.0		497	800C. m-Al-I+Na4SiO4+"510" frit
LD-5	9.20	0.0	850	206	850C. m-Al-I+Na4SiO4+"510" frit
LD-6	9.15	0.0		174	850C. m-Al-I+Na4SiO4+"510" frit
LD-7	8.68	13.5	900	279	900C.m-Al-I+Na4SiO4+"510" frit
LD-8	8.94	7.6		279	900C. m-Al-I+Na4SiO4+"510" frit

Table 6. Summary of leaching tests

The leaching rate of the waste form depends on the stability of both radionuclide-bearing nanocrystallites and their surrounding matrix. HTXRD analyses indicate that at a relatively low sintering temperature, e.g., between 750-800 °C, several crystalline phases appear (Fig. 11). The leaching test result indicates that glass-ceramic waste forms vitrified at 750 °C seem to have the lowest iodine loss during leaching. This may be due to the high content of crystalline quartz (possibly as the embedding matrix) as well as the presence of crystalline lithium silicate. In the case of the glass ceramic sample containing Ag, nanocrystals of AgI are observed to be embedded in crystalline quartz. At a higher vitrification temperature, iodine anions are expected to distribute more uniformly in the resulting waste form, probably "dissolved" in the glass matrix, thus transitioning from nano-scale radionuclide encapsulation to traditional mineral structural incorporation and resulting in less resistant waste forms.

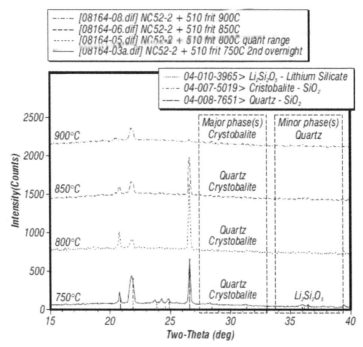

Fig. 11. XRD patterns of glass-ceramic waste forms showing the crystallinity change with increasing vitrification temperatures. Quartz, crystoballite and lithium silicate occur in the 750 °C sample; these phases disappear at 900 °C as the amorphous phase becomes more dominant.

7. Development of high performance adsorbents by molecular design

Adsorption isotherms were simulated using the sorption module of the Materials Studio software suite (Allen and Tildesley, 1987; Frenkel and Smith, 2002). This software suite allows for simulation of sorption isotherms under a wide variety of conditions (temperature and pressure), and has the capability of handling multiple components simultaneously (e.g. I_2 and H_2O). The parameters for a typical isotherm simulation at 363 K use the Metropolis algorithm (Metropolis et al., 1953), with 1.0×10^5 equilibration steps, and 5×10^5 production steps. Electrostatic summations were treated according to the Ewald method (0.001 kcal/mol accuracy) while van der Waals summations were treated with an atom based method (cubic spline cutoff of 12.5Å).

Sorption studies have included γ–alumina (Al_2O_3), and quartz (SiO_2). The isotherms which included simultaneously both I_2 and H_2O were for direct comparison against relative humidity experimental data (Fig. 7). Parameters for γ–Al_2O_3 are Al: σ = 2.898 Å, ε = 0.4916 kcal/mol; O: σ = 3.627 Å, ε =0.0784 kcal/mol; while the parameters for silica: Si: σ = 4.295 Å, ε =0.3000 kcal/mol; O: σ = 3.511 Å, ε = 0.1554 kcal/mol. Since we use Grand Canonical Monte Carlo (GCMC) methods, the sorbate and sorbent models are static, bonding terms need not be included in the force field description.

Because of the large atomic radii of iodine and its ability to become polarized, we have opted to use force-field parameters that mimic this molecular phenomenon. Lennard-Jones 6-12 parameters were used for I_2 (σ = 2.376 Å; ε = 35.56 kcal/mol) (Kornweitz and Levine, 1998); however, the individual iodine atoms of I_2 were artificially charged ($\delta^{\pm}$ = ±0.366 e) (Pasternak et al., 1977) such that molecular neutrality was maintained but a dipole is established (Fig. 12). The result is a sorption analyte which accounts for the molecular polarization of I_2 when it interacts with another atom or surface.

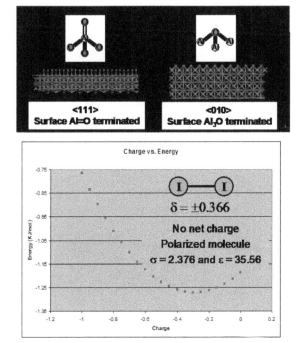

Fig. 12. Model system for iodine sorption on γ-alumina

Comparison of 363K I_2 isotherms between γ–alumina and silica indicate a more significant uptake for the alumina case (Fig. 13). This is consistent with experimental observations that mesoporous silica has a low sorption capability for iodine (Fig. 7). In addition to the maximum uptake capacity, the shape of the isotherm also indicates that the energetics is slightly more favorable in the alumina case. When we compare I_2 uptake versus relative humidity, we can clearly see that increasing the relative humidity reduces the capacity of each material to adsorb molecular I_2 by roughly 50%. This suggests that the I_2 and H_2O are competing for the same surface adsorption sites and that there are no synergistic effects to consider, which was confirmed by our experimental observations (Fig. 7).

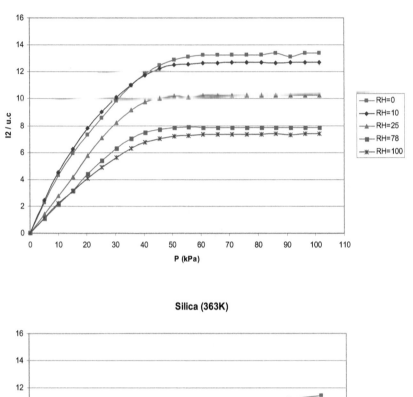

Fig. 13. Molecular dynamic calculations for iodine sorption on alumina and silica surfaces at various relative humidity

The surfaces of these two materials can be terminated structurally in several ways and this could account for the isotherm differences. The first is termination through a single oxygen atom bonded to a single metal (Si or Al) atom (M=O) which produces a surface with a significant regular nanostructures. The second possibility is termination by a single oxygen atom bound to three metal (Si or Al) atoms (M_3O) that produces a very uniform surface with little variation in the local nanostructure. In gamma-alumina the predominant surface termination motif is Al=O and not Al_3O and our model reflects this preference. Whereas in silica, the predominant termination motif is considered Si_3O rather than Si=O and our model also reflects this termination preference. This variation in local surface structure could potentially account for the differences we see in the isotherms; however, further simulations (specifically of the Si=O terminated silica) are necessary to prove this hypothesis.

To investigate the effect of space confinement on gaseous radionuclide adsorption, a GCMC simulation was performed for a nanopore represented by two parallel γ-alumina (Al_2O_3) <111> surfaces. The coordinate perpendicular to the planar walls was changed from 1.5 to 4 nm to simulate different pore sizes. The simulation result for I_2 sorption is s shown in Figure 14. Over the pore size range studied, the pore size from 2.5 to 3.0 nm seems to have optimal sorption performance. This result needs to be confirmed by experiments.

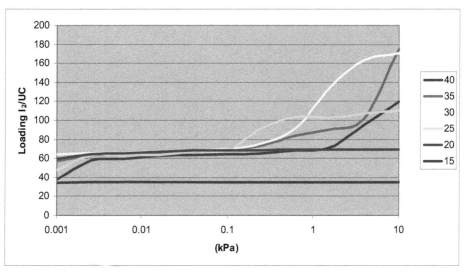

Fig. 14. GCMC adsorption isotherms for iodine sorption onto alumina nanopore surfaces as a function of iodine vapour pressure and pore size. Numbers in the labels are pore sizes in Å.

A GCMC simulation was also performed to investigate the possibility of improving material sorption performance through surface modifications. In the study shown in Figure 15, 60% of oxygen atoms on alumina surface were replaced with F atoms. The simulation result for iodine sorption onto the fluorinated surface is shown in Figure 16. The fluorinated surface appears to have enhanced the attraction of the I_2 molecules to the surface. Based on this result, a technique to graft fluorine functional groups onto nanoporous alumina surfaces is currently under development.

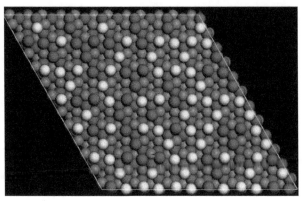

Fig. 15. Plane view of 60% fluorinated alumina surface. Turquoise balls are F, Red = O and Magenta = Al.

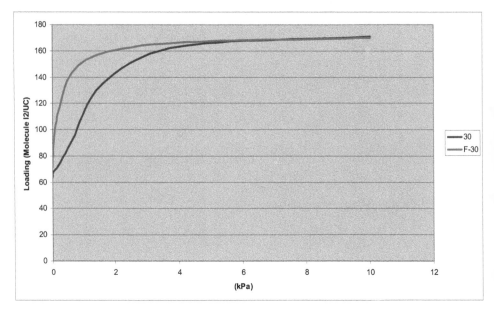

Fig. 16. GCMC simulation result for iodine adsorption onto fluorinated alumina surface. The red curve shows the iodine sorption onto the 60% fluorinated alumina surface.

8. Conclusions

We have proposed a new paradigm for the development of a next generation of high-performance radionuclide adsorbent materials for nuclear waste reprocessing and disposal. Based on this new concept, we have developed a suite of inorganic nanocomposite materials (SNL-NCP) that can effectively entrap various radionuclides, especially for [129]I and [99]Tc. In particular, these materials have high sorption capabilities for iodine gas. After the sorption of radionuclides, these materials can be directly converted into nanostructured waste forms. This new generation of waste forms incorporates radionuclides as nano-scale inclusions in a host matrix and thus effectively relaxes the constraint of crystal structure on waste loadings. Therefore, the new waste forms have an unprecedented flexibility to accommodate a wide range of radionuclides with high waste loadings and low leaching rates.

We have developed a general route for synthesizing nanoporous metal oxides from inexpensive inorganic precursors. A large set of nanocomposite materials have been synthesized and characterized with XRD, BET, and TEM. These materials have been tested for their sorption capabilities for radionuclide I and Re (as an analog to Tc). The results have confirmed that nanoporous Al oxide and its derivatives have high I sorption capabilities due to the combined effects of surface chemistry and nanopore confinement. We have developed a suite of techniques for the fixation of radionuclides in metal oxide nanopores. The key to this fixation is to chemically convert a target radionuclide into a less volatile or soluble form. We have also developed a technique to convert a radionuclide-loaded nanoporous material into a durable glass-ceramic waste form through calcination. We have shown that mixing a radionuclide-loaded getter material with a Na-silicate solution can effectively seal the nanopores in the material, thus enhancing radionuclide retention during waste form formation. Our leaching tests have demonstrated the existence of an optimal vitrification temperature for the enhancement of waste form durability. Our work also indicates that silver may not be needed for I immobilization and encapsulation. Molecular dynamic modeling results help clarify the control of surface structure and surface chemistry on iodine sorption onto SNL-NCP materials and thus provide a guide for future improvements to the materials.

9. Acknowledgment

Sandia is a multiprogram laboratory operated by Sandia Corporation, a Lockheed Martin Company for the United States Department of Energy's National Nuclear Security Administration under contract DE-AC04-94AL85000. This work is supported by DOE Sandia LDRD Program. We thank C. Jeffrey Brinker, Yongliang Xiong, Kathleen Holt, Nathan Ockwig, Mark A. Rodriquez, Denise N. Bencoe, Hernesto Tellez, Jessica Nicole Kruichak, Rigney Turnham, and Andrew Wilson Murphy for their help with the work.

10. References

Allen, M. P. & Tildesley, D. J. (1987) *Computer Simulation of Liquids*. Clarendon Press, Oxford.
Bodansky, D. (2006) Reprocessing spent nuclear fuel, *Physics Today*. December 2006, pp. 80-81.
Ewing, R. C. (2006) The nuclear fuel cycle: A role for mineralogy and geochemistry, *Elements*. Vol. 2, pp. 331-334.
Fortner, J. A.; Kropf, A. J.; Finch, R. J.; Bakel, A. J.; Hash, M. C. & Chamberlain, D. B. (2002) Crystal chemistry of uranium (V) and plutonium (IV) in a titanate ceramic for disposition of surplus fissile material, *J. Nuclear Mater*. Vol. 304, pp. 56-62.
Frenkel, D. & Smith, B. (2002) *Understanding Molecular Simulation: From Algorithms to Applications, 2nd Edition*. Academic Press, San Diego.

Gombert , D. (2007) *Appendixes for Global Nuclear Energy Partnership Integrated Waste Management Strategy Waste Treatment Baseline Study*. Idaho National Laboratory, GENP-WAST-WAST-AI-RT-2007-000324.

Grambow, B. (2006) Nuclear waste glasses - How durable? *Elements*, Vol. 2, pp. 357-364.

Jubin, R. T. (1994) *The Mass Transfer Dynamics of Gaseous Methyl-Iodide Adsorption by Silver-Exchanged Mordenite*. Oak Ridge National Laboratory, ORNL-6853.

Kato, H.; Kato, O. & Tanabe, H. (2002) *Review of Immobilization of Techniques of Radioactive Iodine for Geological Disposal. JAERI-Conf* 2002-004.

Kornweitz, H. & Levine, R. D. (1998) Formation of molecular iodine in high-energy four-center CH3I+CH3I collisions, *Chem. Phys. Lett.* Vol. 294, pp. 153-161.

Lee, J. Y.; Olson, D. H.; Pan, L.; Emge, T. J. & Li, J. (2007) Microporous metal-organic frameworks with high gas sorption and separation capacity, *Advanced Functional Materials*, Vol. 17, 1255-1262.

Metropolis, N.; Rosenbluth, A. W.; Rosenbluth, M. N.; Teller, A. H. & Teller, E. (1953) Equation of state calculations by fast computing machines, *J. Chem. Phys.* Vol. 21, 1087-1092.

NEA OECD (2006) *Advanced Nuclear Fuel Cycles and Radioactive Waste Management*. NEA No. 5990.

Pasternak, A.; Anderson, A. & Leech, J. W. (1977) Bond charge model for lattice-dynamics of iodine, *J. Phys. C: Solid State Phys.* Vol. 10, pp. 3261-3271.

Peters, M. & Ewing, R. C. (2007) A science-based approach to understanding waste form durability in open and closed nuclear fuel cycles, *J. Nuclear Materials*. Vol. 362, pp. 395-401.

Rovnyi, S. I.; Pyatin, N. P. & Istomin, I. A. (2002) Catching of I-129 during processing of spent nuclear fuel from power plants, *Atomic Energy*. Vol. 92, pp. 534-535.

Stephenson, D.J.; Fairchild, C.I.; Buchan, R.M., & Dakins, M.E. (1999) A fiber characterization of the natural zeolite, mordenite: A potential inhalation health hazard, *Aerosol Science and Technology*. Vol. 30, pp. 467-476.

Sudik, A. C.; Côté, A. P.; Wang-Foy, A. G.; O'Keeffe, M. & Yaghi, O. M. (2006) A metal-organic framework with a hierarchical system of pore and tetrahedral building blocks, *Angewandte Chemie*, Vol. 45, 2528-2533.

Wang, Y. & Gao, H. (2006) Compositional and structural control on anion sorption capability of layered double hydroxides (LDHs), *J. Colloid Interface Sci.* Vol. 301, pp. 19-26.

Wang, Y.; Bryan, C.; Gao, H.; Pohl, P.; Brinker, C. J.; Yu, K.; Xu, H.; Yang, Y.; Braterman, P. S. & Xu, Z. (2006) *Potential Applications of Nanostructured Materials in Nuclear Waste Management*. Sandia National Laboratories, Albuquerque, NM. SAND2003-3313.

Wang, Y.; Bryan, C.; Xu, H. & Gao, H. (2003) Nanogeochemistry: Geochemical reactions and mass transfers in nanopores, *Geology*, Vol. 31, 387-390.

Wang, Y.; Gao, H.; Yeredla, R.; Xu, H. & Abrecht, M. (2006) Control of surface functional groups on pertechnetate sorption on activated carbon, *J. Colloid Interface Sci.*, 305, 209-217.

Xu, H. & Wang, Y. (2000) Crystallization sequence and microstructure evolution of Synroc samples crystallized from CaZrTi2O7 melts, *J. Nuclear Mater*. Vol. 279, pp. 100-106.

5

Removal of Selected Benzothiazols with Ozone

Jan Derco[1], Michal Melicher[1] and Angelika Kassai[2]
[1]Institute of Chemical and Environmental Engineering,
Slovak University of Technology, Bratislava,
[2]Water Research Institute, Bratislava,
Slovak Republic

1. Introduction

Benzothiazole derivatives are widely used as industrial chemicals in the leather and wood industries, as bio-corrosion inhibitors in cooling systems, ingredients in anti-freezing agents for automobiles, and mainly as vulcanisation accelerators in rubber production. Correspondingly, these xenobiotic compounds are widely distributed in the environment and they have been detected in industrial wastewaters, as well as in soils, estuarine sediments, and superficial waters (Valdés & Zahor, 2006). They cause environmental concern when released into watercourses (Valdés et al., 2008). These compounds inhibit micro-organisms' activity in conventional biological wastewater treatment systems and most of them are not readily biodegradable (de Wewer & Verachter, 1997; de Wewer, 2007). Moreover, they can be absorbed onto cell membranes, leading to bioaccumulation (Gaja & Knapp, 1998). According to Knapp et al. (1982), 7.0 mg.l^{-1} BT causes a 50% and 54 mg.l^{-1} a 100% inhibition of ammonia oxidation, while nitrate utilization is not affected (de Wever and Verachter, 1997). Tomslino et al. (1966) proved 75% inhibition of ammonium nitrogen oxidation at the concentration of 3 mg.l^{-1} MBT. HOBT causes 100% inhibition of oxidation of ammonium nitrogen at 60 mg.l^{-1} (Hauck, 1972).

Unfortunately, conventional biological wastewater treatment processes are not able to effectively remove such contaminants since these are resistant to biodegradation (Valdés & Zahor, 2006). Thus, the development of efficient treatment/pre-treatment processes is required in order to eliminate their discharge into the aquatic ecosystem. Advanced oxidation processes (AOPs) might be a viable option for the decontamination of biologically recalcitrant wastewaters (Kralchevska et al., 2010). An important group of AOPs are ozone based oxidation procesess, e.g. ozonation at elevated pH, combinations of ozone with UV, hydrogen peroxide etc.

Experimental part of this work was focused on the removal of benzothiazole derivatives by ozone. The results of ozonation trials carried out with the model wastewaters containing single MBT and BT pollutants, the mixture of BT and MBT, the mixtuere of benzothiazole derivatives contained in an industrial wastewater from sulfonamides production as well as with the real wastewater are presented. The conventional ozonation process was investigated with the aim to decrease the toxicity of selected pollutants to microorganisms of activated sludge.

Reaction kinetics was evaluated for two reasons: firstly for quantification of effect of pH on degradation of selected pollutants and secondly, for the estimation of kinetic parameters of main pollutants in model wastewater. The kinetic parameters obtained will be used to design the conditions for ozonation process for treatment of real industrial waste water which contains also other components. Obtained results will be applied for process scale up purposes.

The results of two novel ozonation processes are also presented. Zeolite adsorbent was applied in the adsorption ozonation process. Integrated utilisation of ozone for the degradation of MBT in solutions as well as its degradation after the adsorption on activated sludge and simultaneous disintegration/solubilisation of excess sludge in order to minimise its production was also studied.

2. Mechanisms of ozone oxidation

Ozone, O_3, is a strong oxidant that is able to react through two different reaction mechanisms. Depending on process parameters (pH), the presence of other substances and type/structure of pollutants (Sánchez-Polo et al., 2005) in their molecular (direct reactions of ozone) or hydroxyl radical form (indirect reactions) occur. Molecular ozone is a rather selective oxidant. It can react directly with certain functional groups of organic compounds found in water and wastewaters, such as unsaturated and aromatic hydrocarbons substituted with hydroxyl, methyl and amine groups giving rise to degradation products. Because of this high selectivity, many industrial wastewater treatment oxidation processes can be performed using molecular ozone. On the other hand, ozone decomposes in water to form ·OH radicals which are stronger oxidising agents than ozone itself, thus inducing so-called indirect ozonation. Ozone decomposition in water can be initiated by the hydroxyl anion, HO^-. Indirect ozone oxidation is advanced under alkaline pH conditions (Hoigne & Bader, 1976, 1981).

Major limitations of the ozonation process are the relatively high costs of ozone generation processes coupled with the very short half-life period of ozone. Thus, ozone must always be generated at the site. However, maximum concentration of ozone produced in air or oxygen is approximately 4 to 8%, respectively, which is coupled with the very low (5 to 10%) energy efficiency of the production and the requirement of an absolutely dry input of air or oxygen. Process efficiency is significantly dependent on efficient gas liquid mass transfer, which is quite difficult to achieve due to the low solubility of ozone in aqueous solutions. Schemes of typical equipments used for ozonation and typical operating parameters were presented by Gogate & Pandit (2004).

Total mineralisation with ozone is very expensive. Thus, newly developed processes apply ozone only for the elimination of toxic compounds and/or for partial oxidation of resistant wastewater organic pollutants. Controlled ozonation processes which can inactivate inhibitory compounds in wastewater treatment and improve the biodegradability of recalcitrant organics by altering their chemical structure are applied. The extent of this partial oxidation needs only to be sufficient to facilitate the subsequent biodegradation of converted organic matter.

Another possibility is to use ozone in combination with other techniques such as Ultrasonic/UV radiation, hydrogen peroxide, or other hybrid methods (Gogate & Pandit, 2004; Rodrígues 2009). State of the art of AOPs for wastewater treatment was presented by Poyatos et al. (2010).

The attack by the ·OH radical, in the presence of oxygen, initiates a complex chain of oxidative reactions leading to mineralisation of the organic compound. The exact mechanism of these reactions is still not quite clear. For example, chlorinated organic compounds are first oxidised to intermediates, such as aldehydes and carboxylic acids, and finally to CO_2, H_2O, and the chloride ion. Nitrogen in organic compounds is usually oxidised to nitrate or free N_2, sulphur is oxidised to sulphate (Munter, 2001).

However, the OH radical also reacted with non-targeted substrates, and, as a result, undesirable by-products were produced in some cases. A novel ozonation process using zeolite adsorbents, adsorptive ozonation, was proposed to solve this problem (Fujita et al., 2004). Organic pollutants and ozone can be adsorbed in micropores. As a result, concentrations in micropores become much higher than those in the bulk phase. Therefore, the apparent reaction rate can be increased.

3. Applications of ozone based processes for wastewater treatment

Ozonation can be a suitable technique for the pre-treatment of wastewaters containing low concentrations of toxic or non-biodegradable compounds that cannot be treated with satisfactory results when only traditional, less expensive, biological techniques are applied. In this case, the oxidation process has to be as efficient as possible in order to reduce the costs of ozone addition and energy use. An efficient oxidation process with ozone can be reached by focusing the oxidation with ozone selectively on direct oxidation of toxic pollutants and by minimising ozone loss due to the decay of ozone in water (Boncz et al. 2003).

Applications of ozone techniques for control pollution in full-scale industrial wastewater treatment plants are used in the areas of electroplating wastes, electronic chip manufacture, textiles and petroleum refineries, and treating rubber additive wastewaters (Rice, 1997).

Munter (2001) mentions full scale of ozone applications for decolouration of textile industry wastewaters in several countries and the advantages of ozone and ozone/hydrogen peroxide treatment for the control of microbiological growth in the white water system of the paper machine.

Ozone together with GAC is also efficient considering the solution of environmental problems in the petroleum industry (Munter, 2001).

Because of its high oxidation potential and specific lethality, ozone is the most effective disinfecting agent. When ozone disinfects or breaks down harmful bacteria or pollutants, there are generally no by-products, unlike most other disinfecting agents. Ozone treatment has proven to be very effective for complete removal of colour and for detoxifying of textile washing wastewaters treated for reuse. Acceptance of ozone as a replacement bleaching agent to eliminate the discharge of halogenated effluents from pulp bleaching plants is on the rise.

Improvement of the efficiency of organic pollutants removal was observed in combined systems of ozone with adsorption materials. Zeolite is a natural, low cost, and widely available material, and it makes a suitable candidate for heterogeneous ozonation (Valdes et al., 2009).

Applications of chemical oxidation processes including ozone based degradation processes for complete mineralisation of organics are generally expensive. One attractive potential alternative is the application of these processes by pre-treatment converting the initially resistant (toxic, harmful, persistent) organic compounds into more biodegradable intermediates which can then be mineralised in a biological oxidation process. Many studies have shown that biodegradability of organic components in the waste stream is enhanced when subjected to prior chemical oxidation (Oller et al., 2009). For example, the combination of ozone oxidation followed by biological treatment has been installed in full-scale at a large German industrial chemical complex (Fanchiang et al., 2008).

Ozone also supports other processes. In a newly developed system with ozone-resistant membrane, residual ozone prevents membrane fouling and enables high rate water flux in the micro-filtration process. The use of ozone for pre-treatment prior to adsorption of organics by granular activated carbon has extended the useful life of GAC adsorbers before regeneration is required. Although GAC adsorption alone can reach the same results when removing petroleum related compounds from wash waters, the pre-ozonation step not only extends the operational life of GAC but it does so at a lower total cost.

Contrary to the above-stated drawbacks of ozone, a remarkable advantage of ozone and AOPs in comparison to all conventional chemical and biological processes is that they are totally environmentally friendly. They neither transfer pollutants from one phase to the other (as in adsorption, chemical precipitation, coagulation and volatilisation) nor do they produce massive amounts of sludge (as in activated sludge or coagulation processes).

4. Degradation of benzothiazole derivatives

A Lifetech ozone generator with maximum ozone production of 5 g h^{-1} was used for all experiments carried out in this work. Ozone was prepared from pure oxygen. The values of other parameters and ozonation reactors used for different experiments are briefly specified in the following text.

4.1 Materials and methods

4.1.1 Characterisation of wastewater

Model wastewater contained some or all of the following compounds: 2-mercaptobenzothiazole (MBT), benzothiazole (BT), 2-hydroxybezothiazole (OHBT),

2-aminobenzotiazole (ABT), and aniline (AN) as selected organic pollutants of an industrial wastewater depending on experimental trials. All chemicals were provided by Merck (≥ 95.5% purity).

4.1.2 Experimental equipment and procedures

Ozonation experiments were carried out in a lab-scale bubble ozonation column, jet loop ozonation reactor and a completely mixed ozonation reactor.

The ozonation glass columns have a diameter of 0.04 meter and height of 1.7 m. Ozone was produced from oxygen by an ozone generator with maximum production of 5 g.h^{-1} of ozone. The mixture of oxygen and ozone was injected at the bottom through porous air diffusers with a constant flow rate of 40 l.h^{-1}.

The ozonation column was filled with 1 litre of wastewater. The experiments were carried out with the synthetic wastewater containing 1 g.l^{-1} of MBT. The system was operated in the batch mode.

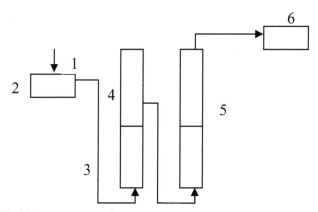

Fig. 1. Scheme of bubble ozonation column
1 - feed of oxygen, 2 - ozone generator, 3 - feed of ozone, 4 - ozonation column with wastewater, 5 - ozonation column with a solution of KI, 6 – destruction of ozone

A scheme of the jet loop ozonation equipment is shown in Figure 2. The system was operated in the batch mode with regard to the wastewater samples. The samples were added into a jet loop ozonation reactor at the beginning of the trials.

A mixture of O_3 and O_2 was injected into a wastewater sample through a Venturi ejector. At the same time, the ejector sucked the mixture of O_3 and O_2 from the reactor headspace. This, together with external circulation, should improve the efficiency of ozone utilisation in the ozonation reactor. The outlet gas mixture was conducted into a bubble column through a fine-bubble porous distribution element. The bubble column had 0.04 m in diameter and 1.7 m in height. The column was filled with a solution of potassium iodide. The excess ozone destruction was carried out in this column. Similarly to the ozonation reactor, effective volume of the bubble column was 1.0 dm^3.

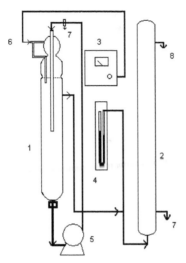

Fig. 2. Scheme of jet loop ozonation equipment
1 – ozonation jet loop reactor, 2 – destruction of excess O_3, 3 – ozone generator,
4 – manometer, 5 – pump, 6 – inlet of the mixture of O_3 and O_2, 7 – sampling, 8 – gas outlet

The continuous flow of oxygen of 20 l h^{-1} was applied for the generation of ozone. Ozonation trials were carried out at 30% of the ozone generator's power maximum.

Some experiments were carried out in a batch stirred ozonation reactor (Figure 3) with the reactor volume of 200 ml. A magnetic stirrer was used to homogenise the reaction mixture of 100 ml of model wastewater and zeolite (grain size d = 0.315 to 0,400 mm).

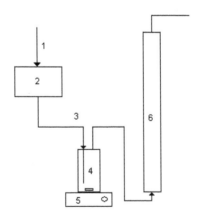

Fig. 3. Scheme of mechanically stirred ozonation reactor
1 - oxygen, 2 – ozone generator, 3 a mixture of O_2 and O_3, 4 – ozonation reactor
5 – magnetic stirrer, 6 – residual ozone destruction in the KI solution

The same ozone generator (Lifetech) as in previous experiments using the recirculation reactor (Figure 2) and column apparatus (Figure 1) was used. The flow of oxygen to the ozone generator was maintained at 40 l.h^{-1}.

The influence of raw and ozonated wastewater samples on the activated sludge microorganisms activity, respirometric measurements (Spanjer et al, 2000) and a nitrification kinetic test were carried out with activated sludge sampled at the industrial wastewater treatment plant.

Specific ozone consumption was evaluated based on the balance of ozone concentration in the influent and effluent gas. A UV detector Life ODU 100 was used to analyse the ozone gas content.

4.1.3 Analytical procedures

Analytical control of raw wastewater and monitoring of the treatment procedures included pH, and COD (Chemical Oxygen Demand) determination (ISO 6060, 1989; Greenberg et al, 2005). TOC was measured with analyser Shimdzu TOC-V$_{CPH/CPN}$ (USA). MBT, BT and their derivatives were analysed using a Hewlett Packard Liquid chromatograph series II 1090 with a DAD detector. The direct injection method was applied using linear gradient of RP-HPLC with a UV-DAD detector on column C18 (Merck).

4.1.4 Mathematical treatment of experimental data

Experimental data were fitted by zero (Eq. (1)), first (Eq. (2)), and second (Eq. (3)) order reaction kinetic models. Besides these single power law models, a combined – "two components" (TCM) kinetic model (Eq. (4)) with the first order reaction kinetics was applied to proportions of easily and slowly oxidizable organics. For a batch reaction system, under the assumption of a constant reaction volume, the following relationships were obtained for COD (in general for substrate S)

$$COD_t = COD_0 - k_0 t \tag{1}$$

$$COD_t = COD_0 \exp(-k_1 t) \tag{2}$$

$$COD_t = \frac{COD_0}{(1 + COD_0 k_2 t)} \tag{3}$$

$$COD_t = COD_{EO,t} + COD_{SO,t} = \alpha COD_0 \exp(-k_{EO} t) + (1 - \alpha) COD_0 \exp(-k_{SO} t) \tag{4}$$

where $COD_t /$ (g m^{-3}) denotes the value of COD in wastewater in time t, $COD_0 /$ (g m^{-3}) the initial value of COD in wastewater, $k_0 /$ (g m^{-3} h^{-1}), $k_1 /$ h^{-1}, $k_2 /$ (g^{-1} m^3 h^{-1}) the rate constants for the zero, first and second order kinetics, respectively, $COD_{EO,t} /$ (g m^{-3}) the value of easily oxidizable organics, $COD_{SO,t} /$ (g m^{-3}) the value of slowly oxidizable organics, $k_{EO} /$ h^{-1} the first order reaction rate constant for easily oxidizable organics, $k_{SO} /$ h^{-1} the first order reaction rate constant for slowly oxidizable organics, and a the portion of easily oxidizable organics.

Parameter values of the applied kinetic models were calculated by the grid search optimisation procedure. The residual sum of squares (S_r^2) between the observed values and the values given by the model, divided by its number of degrees of freedom v (the number of observations less the number of parameters estimated) was used as the objective function (S_r^2).

COD reduction by partial oxidation can be done by the following equations (Fiehn et al, 1998):

COD removal in total:

$$\alpha COD_{oxi} = 1 - \frac{COD_t}{COD_0} \tag{5}$$

COD removal due to mineralisation:

$$\alpha COD_{min} = 1 - \frac{DOC_t}{DOC_0} \tag{6}$$

COD removal due to partial mineralisation:

$$\alpha COD_{partoxi} = \alpha COD_{oxi} - \alpha COD_{min} \tag{7}$$

Degree of effectiveness of partial oxidation:

$$\mu COD_{partoxi} = \frac{\alpha COD_{partoxi}}{\alpha COD_{oxi}} \tag{8}$$

Ozone transferred in the reactor was calculated by macroscopic mass balance across the reactor:

$$O_{3,trans} = Q_g \int_0^t \frac{O_{3,in} - O_{3,out}}{V_R} dt \tag{9}$$

where $O_{3,trans}$ denotes the ozone transferred [g.l^{-1}], Q_g the gas flow rate [Nm^{-3}. min^{-1}], $O_{3,in}$ the ozone concentration in the gas at the inlet [g.Nm^{-3}], $O_{3,out}$ the ozone concentration in the gas at the outlet [g.Nm^{-3}] and V_R is the volume of the reactor [l].

4.2 Ozonation of 2-mercaptobenzothiazole

Ozonation trial with model wastewater containing 6.6 mmol.l^{-1} of 2-Mercaptobenzothiazole (MBT) was performed in a bubble ozonation column at the oxygen flow $Q_{O2} = 30$ l.h^{-1}. Performance of the ozone generator was maintained at 80% of the maximum value. Ozone concentration in the influent gas was 99.4 g.Nm^{-3}. Transferred ozone followed linear dependence on the reaction time ($R^2 = 0.9997$). Average rate of ozone transfer was app. 3.0 g.h^{-1}.

Fig. 4 illustrates the decline of MBT and the evolution of Benzothiazole (BT) and Bezothiazole-2-sulfonate (BTS) as well as the time course of the sum of identified derivatives during ozonation of model wastewater containing MBT.

Total removal of MBT in 15 minutes of ozonation can be seen in this figure. Degradation of MBT followed the first order reaction rate (Eq. 2) with the reaction rate constant of 0.1365 h^{-1} ($R^2 = 0.9264$). On the other hand, formation of intermediates, mainly BT and BTS, is evident from the figure. In addition to the BT derivatives presented in Fig. 4, the presence of Hydroxybezothiazole (OHBT) was observed with the maximum content after 10 minutes of ozonation and corresponded to about 4.0% of total BT derivatives at this stage of the trial. No presence of OHBT was observed after 30 minutes of ozonation. Only about 41.6% of organic carbon was oxidised (Eq. 5) during ozonation. About 40.3% of oxidised organic carbon was mineralised (Eq. 6) and the remaining 59.7% of oxidised organics were only partially oxidised (Eq. 7).

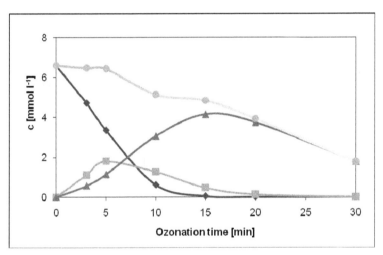

Fig. 4. Dependencies of BT derivatives' content on the ozonation time
MBT, ■ BT, ▲ BTS, ● Sum of BT derivates

The presented results indicate that although the target pollutant was completely removed, intermediates and final oxidation products should be analysed with regard to possible inhibition of subsequent biological processes or even direct discharge into the aquatic environment.

4.3 Ozonation of benzothiazole

The next set of ozonation trials was performed with model wastewater containing BT. Ozonation was carried out in the bubble ozonation column at the oxygen flow Q_{O2} = 60 l.h^{-1}. Performance of the ozone generator was maintained at 80% of its maximum value. Dependencies and efficiency of BT removal on the ozone supplied are plotted in Figure 5.

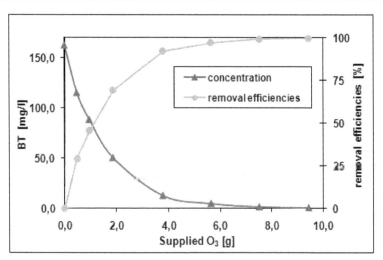

Fig. 5. BT removal dependencies on ozone supplied during ozonation of model wastewater

The concentration of BT dropped to 0.01 after 80 minutes of the process (about 8 g.l^{-1} of ozone supplied). The best fit of BT experimental data was obtained by the first-order kinetic model (Eq. 2) with the reaction rate constant of 6.14 10^{-2} h^{-1} (R^2 = 0.9989).

The time dependencies of COD and TOC removal during ozonation are presented in Figure 6.

The highest removals of TOC and COD were observed during the first 40 minutes of ozonation (ozone dose of app. 4.0 g.l^{-1}). This corresponds also to the highest BT removal rate (Figure 5).

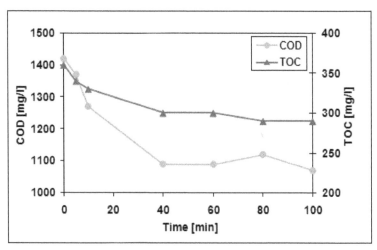

Fig. 6. COD and TOC removal dependencies on ozonation time

On the other hand, the values of COD and TOC are too high when realising that the model wastewater contains only one component. Only about 25% removal of TOC was observed (Eq. 5) during the process. About 70.9% of oxidised organic carbon was mineralised (Eq. 6) and the remaining 19.1% of oxidised organics were partially oxidised (Eq. 7).

The results of ozonation assay carried out in a completely mixed ozonation reactor with zeolite addition to synthetic wastewater containing BT are presented in this part.

Natural zeolitic tuff from the deposit in Nižný Hrabovec, Slovak Republic, with the granularity from 0.315 to 0.4 mm was used. This type of zeolite consists mostly of clinoptilolite (from 40 to 70%), quartz (2 to 5%), α cristobalite (6 to 10%), feldspar (8 to 10%) and volcanic glass from 13 to 30% (Lukáč et al., 2005).

The dependencies of BT concentration values and the efficiency of its removal as a function of zeolite dose are plotted in Figure 7. Ozonation time was 30 minutes and performance of ozone generator was 50% of its maximum power.

The efficiency of BT removal increased by 10% to 20% in the presence of zeolite in comparison with the ozonation of BT in its absence. However, variability of the zeolite dose did not influence the efficiency of BT removal significantly.

The influence of ozonation time on the BT removal efficiency is presented in Figure 8. Performance of the ozone generator was maintained at 50% of the maximum power and 0.5 g of zeolite was applied in this set of experiments.

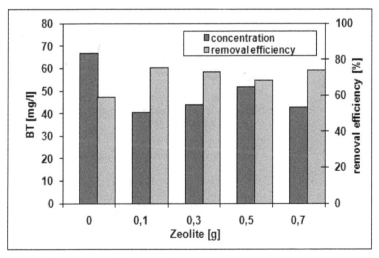

Fig. 7. Effect of zeolite dose on BT removal by ozonation

Significant positive influence of ozonation time on the efficiency of BT removal is obvious from Figure 8. The efficiency of BT removal exceeded 96% after 60 minutes of the process.

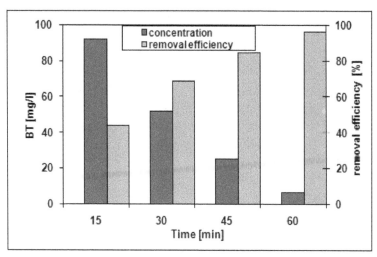

Fig. 8. Influence of ozonation time on BT removal efficiency

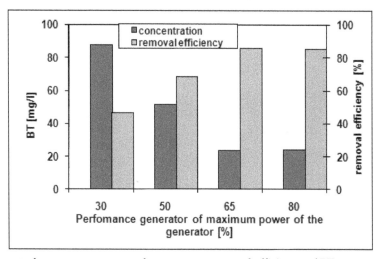

Fig. 9. Impact of ozone generator performance on removal efficiency of BT

On the other hand, the efficiency of BT removal increased by 20% in the presence of zeolite when comparing the results of both assays (Figures 7 and 9) performed at the ozone generator power of 50% of the maximum and 30 minutes of ozonation. As it can be seen from Figure 9, an increase of the ozone generator performance to over 65% of its maximum power did not lead to further increase of the BT removal efficiency.

The influence of the ozone generator performance on the BT removal in the presence of zeolite is presented in Figure 9. Ozonation time was 30 min and the dose of zeolite was 0.5 g. The increase of the ozone generator power from 50 to 65% of its maximum power resulted in an increase of the BT removal efficiency by 18%.

4.4 Ozonation of benzothiazole derivatives

A mixture of BT and MBT was also treated. Ozonation was carried out in the bubble ozonation column at the oxygen flow Q_{O2} = 60 l.h^{-1}. Performance of the ozone generator was maintained at 80% of its maximum value.

MBT and BT removals during ozonation are presented in Figure 10. The pH value in the model wastewater containing BT and MBT was adjusted to 11.76 due to low solubility of MBT in water at lower pH.

From Figure 10 it follows that the rate of MBT degradation is higher in comparison to the BT removal. The effectiveness of MBT removal was more than 99% after 10 minutes of ozonation. On the other hand, BT removal efficiency was only 21.5% after 10 minutes of the process. The decrease of COD (about 13%) and TOC (about 7%) was lower in comparison to the decrease of the BT and MBT content. Ozone dose with time was linear with ozonation time and was app. 0.1 g.min^{-1}.

Kinetic parameter and correlation coefficients are given in Table 1. Time decline of MBT conformed to the first order kinetic model. The best fits of BT removal from the model wastewater containing BT and MBT was obtained with the kinetic model of the zero order (Table 1).

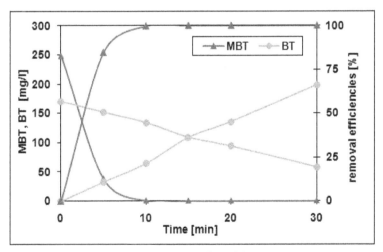

Fig. 10. BT and MBT removal during ozonation of model wastewater

Model	Parameter, objective function	Removal of	
		MBT	BT
Eq. (1)	$k_0/(g\ m^{-3}\ h^{-1})$		3.8412
	R^2		0.9969
Eq. (2)	k_1/h^{-1}	0.382	
	R^2	0.9994	

Table 1. Kinetic parameters and correlation coefficients during ozonation of model wastewater containing BT and MBT

Time dependencies of BT removal during ozonation of BT and BT/MBT model wastewaters are presented in Figure 11.

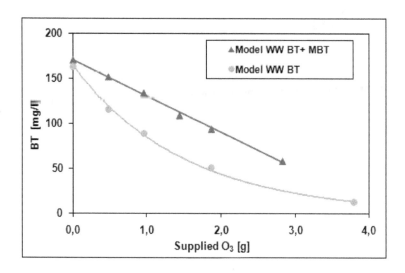

Fig. 11. BT removal during ozonation of model wastewaters

As it can be seen in Figure 11, kinetics of BT removal is lower during ozonation of the model wastewater containing BT and MBT (3.85 mg.l^{-1}. min^{-1}) in comparison with BT model wastewater (5.60 mg.l^{-1}. min^{-1}). This can be explained by the transformation of MBT to BT (Figure 4) during ozonation.

The results of model wastewater ozonation containing MBT, BT, OHBT, 2-Aminobenzthiazole (ABT), BTS and Aniline (AN), as the main organic pollutants of industrial wastewater produced in the production of N-cyclohexyl-2-benzothiazol-sulfenamide, are presented in the next part. Brief characterisation of the industrial wastewater was reported by Derco at al. (2001). An ozonation jet loop reactor with external recirculation of the reaction mixture was applied in these experiments. As it was already mentioned, the mixture of oxygen and ozone was injected into a wastewater sample through a Venturi ejector. At the same time, the ejector sucked the gas mixture from the reactor headspace. This, together with external circulation, should improve the efficiency of ozone utilisation in the ozonation reactor.

Continuous flow of oxygen of 20 dm^3 h^{-1} was applied for the generation of ozone. Ozonation trials were carried out at 30% of the ozone generator's power maximum. External circulation of the reaction mixture was equal to the previous experiments performed in the jet loop reactor.

Dependencies of MBT, BT and BTS removal on the ozone supplied at pH = 8.5 and 5.8 are presented in Figure 12 and 13, respectively.

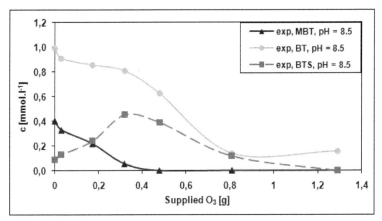

Fig. 12. Dependencies of MBT (▲), BT (•) and BTS (■) removal on ozone supplied at pH 8.5

Higher removal rate of MBT was observed at higher pH value. A 99.5% MBT removal efficiency was obtained after 40 minutes of the process (0.48 g.l^{-1} of ozone supplied/transferred). On the other hand, only a 75.7% MBT removal efficiency was measured at lower pH value.

Significantly lower decline of BT content are obvious from Figures. 12 and 13. Only a 36.6% BT removal efficiency at higher pH and a 0.97% BT removal efficiency at lower pH value were observed after 40 minutes of ozonation. However, removal rate of BT increases after the depletion of MBT and final observed removal efficiency at higher pH was 85.9%. On the other hand, concentration of BTS increases with the decline of MBT and decreases after the MBT removal (Figure 13), i.e. maximum BTS concentration corresponds to minimum MBT concentration.

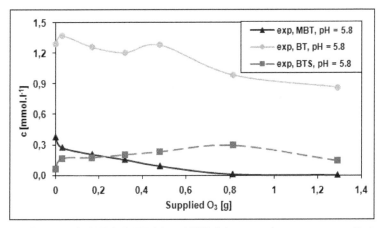

Fig. 13. Dependencies of MBT (▲), BT (•) and BTS (■) removal on ozone supplied at pH 5.8

Higher pH value is favourable also for OHBT and ABT removals. Similarly to BT and BTS removals also removal rates of OHBT and ABT significantly increased after the MBT removal. The final observed removal efficiencies of OHBT and ABT were 87.5% and 84.9% respectively.

Results of the ozonation of industrial wastewater produced by the manufacture of sulphenamides are presented in the next part. BT and MBT were the main pollutants; OHBT, ABT, MBT and AN were also present in the investigated wastewater.

The ozonation jet loop reactor with external recirculation of the reaction mixture was applied also in these experiments. Continuous flow of oxygen of 60 l. h-1 was applied for the generation of ozone. Ozonation trials were carried out at 30% of the ozone generator's power maximum. External circulation of the reaction mixture was maintained at 0.5 l. min-1 by a membrane pump.

MBT removal dependencies on the ozonation time at different pH values are presented in Figure 14. The best fit of MBT experimental data was obtained by the second order kinetic model. The rate constant values and the values of S_r^2 relevant to the results of the MBT removal obtained by this model are given in Tables 2 and 3.

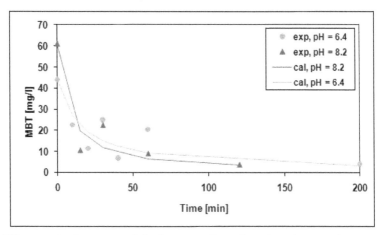

Fig. 14. MBT concentration removal during ozonation of wastewater samples at different pH

The highest removal rate of MBT with ozone was observed during the first 10 - 15 min of ozonation. Corresponding transferred ozone was 0.38 to 0.55 g.l-1 O$_3$. The conversion of MBT reached about 60% to 83% after this period of ozonation.

Model	Parameter, objective function	Removal of		
		MBT	BT	AN
Eq. (2)	k_1/h^{-1}	-	-	5.0×10^{-2}
	S_r^2	-	-	1.8×10^1
Eq. (3)	$k_2/(g\ m^{-3}\ h^{-1})$	1.4×10^{-3}	4.0×10^{-5}	–
	S_r^2	6.7×10^1	3.5×10^2	–

Table 2. Kinetic parameters and statistical values – ozonation at pH = 6.4

Figure 15 shows the time courses of BT removal during ozonation of wastewater samples performed at different pH values. Similarly to MBT, the best description of experimental data for BT removal was obtained by the second order kinetic model. The values of kinetic parameters and relevant S_r^2 for the BT experimental data obtained by the applied kinetic model are also summarised in Tables 2 and 3.

Model	Parameter, objective function	Removal of		
		MBT	BT	AN
Eq. (2)	k_1/h^{-1}	-	-	7.3×10^{-2}
	S_r^2	-	-	1.8×10^{-1}
Eq. (3)	$k_2/(g\ m^{-3}\ h^{-1})$	2.3×10^{-3}	7.6×10^{-5}	–
	S_r^2	6.8×10^2	1.8×10^3	–

Table 3. Kinetic parameters and statistical values – ozonation at pH = 8.2

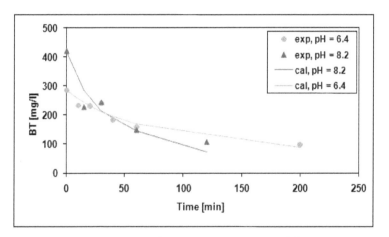

Fig. 15. BT removal during ozonation of wastewater samples at different pH

Efficiency of the BT removal achieved about 19% to 46% after the above mentioned period of ozonation. Higher value of pH was more favourable both for MBT and BT removal.

Similar results were presented by Valdés et al. (2003) who investigated the importance of pH for indirect reactions of ozone with BT. According to their results, the influence of pH on the reaction rate is significant particularly at low pH values. However, only small differences in the BT removal carried out at pH = 7 and pH = 9 were observed by the authors.

Very fast decline of the AN content, with the treatment efficiency of above 90%, was also achieved at this stage of the process (Figure 16). Slightly higher removal efficiency rate of AN was observed also at higher pH value with an about 68% removal efficiency during the first 15 minutes of ozonation. The final observed treatment efficiency of aniline was about 99%. The AN experimental data conformed to the first order kinetic model. The statistical and kinetic parameter values are summarised in Tables 2 and 3.

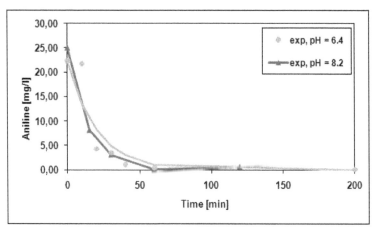

Fig. 16. Aniline removal dependencies on the reaction time during ozonation of wastewater samples at different pH

Concentration values of ABT and OHBT remained more or less stable during the ozonation trials but their concentrations were lower by more than one order in comparison with the previous compounds.

COD removals during ozonation performed at different pH are presented in Figure 17. It is obvious that the removal efficiencies of organics are significantly lower when compared with those of individual organic pollutants (Figures 14 to 16). However, the highest COD removal was observed within the first 15 – 20 minutes of ozonation (transferred ozone 0.38 to 0.55 g.l^{-1} O$_3$) similarly to other monitored pollutants.

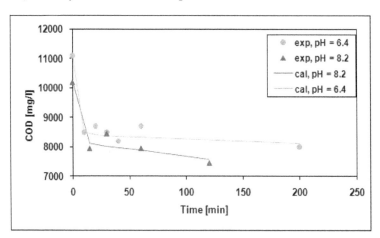

Fig. 17. COD removal dependencies on the reaction time during ozonation of wastewater samples at different pH

Slightly higher removal rate of COD was observed at the lower pH value. Kinetics of COD removal follows the ´two components´ model (Eq. 4) with a 24% portion of easily oxidizable compounds in the wastewater at the lower pH value (k_{EO} = 0.32 h^{-1}, k_{SO} = 1.78 10^{-4} h^{-1}, R^2 = 0.9562).

Figures 18 and 19 show the portions of partially oxidised COD versus mineralised COD measured during ozonation at both pH values.

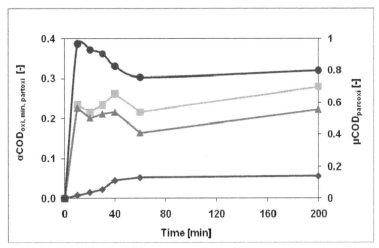

Fig. 18. Influence of ozonation time (pH 6.4) on COD partial oxidation and mineralisation (■) aCOD$_{oxi}$, (▲) aCOD$_{min}$, (◆) aCOD$_{parcoxi}$, (●) μCOD$_{parcoxi}$

It is obvious that the highest degree and efficiency of partial oxidation was achieved within 10 to 15 minutes of ozonation at both pH values. However, the lower pH value is more favourable for partial oxidation.

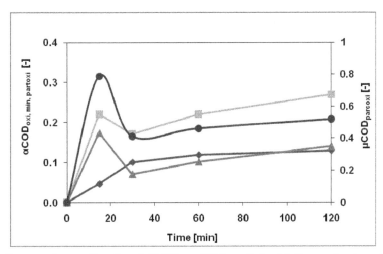

Fig. 19. Influence of ozonation time (pH 8.2) on COD partial oxidation and mineralisation (■) aCOD$_{oxi}$, (▲) aCOD$_{min}$, (◆) aCOD$_{parcoxi}$, (●) μCOD$_{parcoxi}$

Batch kinetic tests were performed with raw and ozonated wastewater samples and non-acclimated activated sludge sampled from a refinery WWTP. From Figure 20 follows a higher BT removal efficiency for ozonated WW sample (70.8%) in comparison with untreated wastewater sample (63.3%).

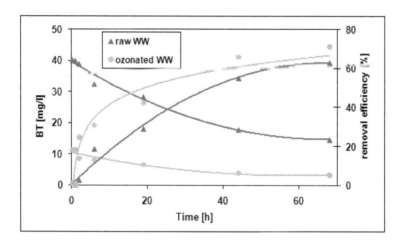

Fig. 20. BT removal during kinetic tests carried out using raw and ozonated wastewater

Complete removal of AN was observed in both the kinetic tests. The concentration values of HOBT and ABT were more or less constant during the tests.

Influence of the ozonated wastewater sample on the nitrification activity of activated sludge microorganisms was studied. Batch tests were carried out with activated sludge sampled from an industrial WWTP. Practically complete inhibition of the second nitrification step is characteristic for the applied activated sludge (Buday et al., 2000). Theis sample of industrial wastewater was used for reference measurements. Relatively low inhibition (up to 4.0%) of the ozonated wastewater sample in the first nitrification step was observed.

4.5 Integrated ozonation process

An ozonation jet loop reactor with external circulation of the reaction mixture (Fig. 2) was used also for this set of experiments. Ozonation trials were carried out at the oxygen flow rate of 10 dm³ h⁻¹ and at 30% of the maximum ozone generator's power.

Results of the integrated processes of excess activated sludge solubilisation and simultaneous degradation of dissolved pollutants with ozone are presented in Figures 21 and 22.

Cell membranes of the activated sludge are destroyed by ozone and the intercellular material is released into the liquid phase (Cui & Jahng, 2004). Sludge ozonation is referenced as one of the most cost effective technologies with the highest disintegration capability (Müller, 2000; Park et al., 2003). Furthermore, ozonated sludge can be effectively utilised as an additional carbon source in a biological nitrogen removal process saving costs on an external carbon source (Ahn et al., 2002).

Impacts of non-ozonated and ozonated (20 min of ozonation) samples of MBT on the specific oxygen uptake rate of activated sludge microorganisms are presented in Fig. 21. Respirometric measurements were performed according to Spanjer et al (2000).

It is obvious, from the results, that untreated MBT exhibits an inhibition effect on the specific oxygen uptake rate already at very low concentration values in terms of COD. On the other hand, after 10 minutes of ozonation, the sample showed a stimulating effect on the respiration activities of activated sludge microorganisms oxygen uptake of up to 50 mg.l⁻¹ of organics in the COD term. Efficiency of the MBT removal was 85.5% after 20 minutes of ozonation.

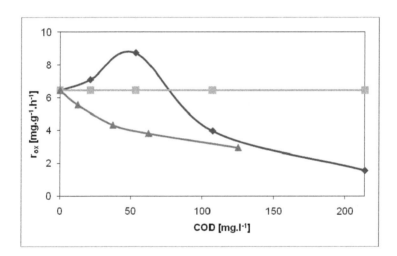

Fig. 21. Influence of ozonated (20 min) and non-ozonated MBT on the oxygen uptake rate
■ $r_{end,}$ ◆ $r_{ox,treated,}$ ▲ $r_{ox,untreated}$

Influence of the liquid part of solubilised activated sludge by ozone performed in the presence of MBT on the specific oxygen uptake rate of activated sludge microorganisms is presented in Figure 22.

Removal efficiency of MBT after 20 minutes of ozonation of the mixture of activated sludge and MBT was 99%. MBT was completely removed from the reaction mixture after 60 minutes of ozonation. Although the efficiencies of the MBT removal are very close for the 20 and 60 minutes ozonation, the impact of these ozonated samples on the respiration activity of activated sludge microorganisms is significantly different.

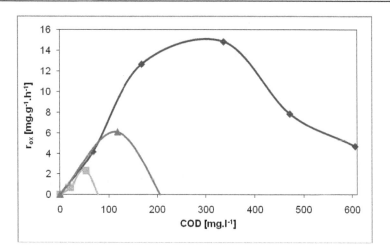

Fig. 22. Influence of ozonated AS and/or MBT on the oxygen uptake
♦ ozonation of AS +MBT – 60 min, ■ ozonation of MBT – 20 min, ▲ ozonation of
AS + MBT – 20 min

5. Conclusions

Each of the various physical, chemical and biological processes developed for wastewater treatment has its own limitations in its applicability, effectiveness or cost.

Ozone is a very powerful oxidising agent that can react with most species containing multiple bonds, but not with singly bounded compounds at high rates. At lower pH, direct selective oxidation predominates, while indirect reaction mechanisms dominate at higher pH values.

The main drawbacks of the ozonation process are relatively high operational costs related to the energy necessary for ozone preparation. Thus, ozonation alone is expensive in comparison with biological or some chemical oxidation processes due to the requirement of a high ozone dose needed for the complete degradation of organics. On the other hand, a very important advantage of ozone-based oxidation processes is that no chemical sludge is produced and thus there are no associated disposal costs in comparison with other chemical methods such as coagulation or precipitation.

Feasibility of the ozonation process to remove selected benzothiazols and the facilitation of subsequent biological wastewater treatment were studied. Results of ozonation trials carried out with the model wastewaters containing single MBT and BT pollutants, a mixture of BT and MBT, mixture of benzothiazole derivatives contained in an industrial wastewater from sulphenamides production and with real wastewater are presented. Total removal rate of MBT occurred within 10 to 15 minutes of ozonation (transferred O_3 of app. 0.5 g.l^{-1}) independently on the model wastewater composition. Formation of intermediates, mainly BT, BTS, and OHBT was observed. The best fit of MBT and BT experimental data was obtained by the first-order kinetic model for single model wastewater containing single pollutants.

Lower removal rate of BT was observed in comparison with that of MBT, particularly in the presence of MBT. This can be explained by the transformation of MBT to BT (Figure 4) during ozonation. BT experimental data were well fitted by the zero-order kinetic model in this case.

Experimental results indicate that a combination of ozonation with zeolite enhances the BT removal efficiency and consequently enables a decrease of operational costs. BT removal efficiency increased by 10 to 20% in the presence of a zeolite.

Significant influence of MBT on the BT removal was observed when the model wastewater containing the main BT derivatives present in wastewater from sulphenamides production was ozonised. Removal rate of BT increased after the depletion of MBT. On the other hand, concentration of BTS increased with the decline of MBT and decreased after the MBT removal. Similarly to BT and BTS removals also removal rates of OHBT and ABT significantly increased after the MBT removal. Higher pH value was favourable for MBT, BT, OHBT and ABT removals.

Higher pH value was also favourable for MBT and BT removal from the industrial wastewater. The best description of experimental data BT and MBT removal were obtained by the second order kinetic model. Concentration values of ABT and OHBT remained more or less stable during the ozonation trials but their concentrations were lower by more than one order in comparison to the previous compounds. The kinetics of COD removal follows the 'two components' model with a 24% portion of easily oxidizable compounds in wastewater at lower pH. Similarly to COD, the lower pH value is more favourable also for partial oxidation.

The combination of ozonation as a pre-treatment process followed by biological oxidation integrates the benefit of low cost of biological treatment with that of ozonation. The aim of controlled ozonation is only partial oxidation of low-biodegradable complex organics. Higher BT removal efficiency for ozonated wastewater sample than for an untreated one was observed in a batch kinetic test. Relatively low inhibition of the ozonated wastewater sample in the first nitrification step was observed.

Ozonation might not always be able to enhance the biodegradability of complex industrial wastewaters. Experimental results of benzothiazole derivative degradation by ozone confirm the necessity of evaluating the effect of ozone on the characteristics of real complex wastewater in order to determine the appropriate ozonation conditions. Experimental results illustrate that prolonged ozonation time significantly decreases toxicity and enhances biodegradability by a continuing transformation of the initial intermediates of the ozonation process.

The results indicate that another possibility of improving the cost efficiency of the ozonation process is to integrate treatment targets, e.g. excess activated sludge minimisation simultaneously with the degradation and removal of micropollutants.

6. References

Ahn K. H., Yeom I. T., Park K. Y. Maeng S. K., Lee Y. H. & Song K. G., (2002). Reduction of sludge by ozone treatment and production of carbon source for denitrification. *Wat Sci. Technol.* 46, (11-12) pp. 121-125.

Boncz M. A., Bruning H. & Rulkens, W. H., (2003). Innovative reactor technology for selective oxidation of toxic organic pollutants in wastewater by ozone. *Water Science and Technology* 47, (10) pp. 17-24.

Buday J., Halász L., Drtil M., Bodík I., Németh P. & Buday M., (2000). Nitrogen removal from wastewater of chemical company Duslo. *Wat. Sci. Tech.*, 41, (9) pp. 259-264.

Cui R. & Jahng D., (2004). Nitrogen control in AO process with recirculation of solubilized excess sludge. *Water Res.* 38, pp. 1159-1172.

de Wewer H. & Verachter H., (1997). Biodegradation and toxicity of benzothiazoles. *Wat. Res.* 31 (1997), pp. 2673-2684.

de Wewer H., Weiss S., Reemtsma T., Muller J., Knepper T, Rorden O. , Gonzalez S., Barcelo D. & Hernando M. D., (2007). Comparison of sulphonated and other micropollutants removal in membrane bioreactor and conventional wastewater treatment. *Wat. Res.* 41, pp. 935-945.

Derco J., Gulyásová A., Králik M. & Mrafková L., (2001). Treatment of an industrial wastewater by ozonation. *Petroleum and Coal*, 43 (2) 92-97.

Fanchiang J.-M., Tseng D.-H., Guo G.-L. & Chen H-J., (2008). Ozonation of complex industrial park wastewater: effects on the change of wastewater characteristics. *J. Chem. Technol. Biotechnol.* 84, pp. 1007-1014.

Fiehn O., Wegener G., Jochimsen J. & Jekel M., (1998). Analysis of the ozonation of 2-mercaptobenzothiazole in water and tannery wastewater using sum parameters, liquid- and gas chromatography and capillary electrophoresis. *Wat. Res.*, 32, pp. 1075-1084.

Fujita H., Izumi Y., Sagehashi M., Fuji T. & Sakoda A., (2004). Adsorption and decomposition of water-dissolved ozone on high silica zeolites. *Water Research*, 38, 159-165.

Gaja M. A. & Knapp J. S., (1998). Removal of 2-mercaptobenythiazole by activated sludge: a cautionary note. *Wat. Res.* 32, pp. 3786 – 3789.

Gogate P. R. & Pandit A. B., (2004). A review of imperative technologies for wastewater treatment I: oxidation technologies at ambient conditions. *Advances in Environmental Research* 8, pp. 501-551.

Greenberg A. E., Clesceri L. S. & Eaton A. Eds., (2005). *Standard Methods for the Examination of Water and Wastewater.* 22nd Edition. Washington, DC.: American Public Health Association.

Hauck R. D., (1972). *Synthetic slow release fertilizers and fertilizer amendments, pp 633-690. In Organic Chemicals in the Soil Environment.* Edited by Goring C. A. I. and Hamaker J. W.

Hoigné J., Bader H., (1986). *Determination of ozone and chlorine dioxide in water by the indigo method, analytical aspects of ozone treatment of water and wastewater.* Lewis Publisher, Michigan.

ISO 6060, (1989). *Water quality – Determination of chemical oxygen demand.* Geneve, Switzerland: International Organisation for Standardisation. 1989.

Knapp J. S., Callez A. G. & Minpriye J., (1982). The microbial degradation of morpholine. *J. Appl. Bacteriol.*, 40, pp. 5-13.

Kralchevska R., Premru, A., Tišler T., Milanova, M., Todorovsky D. & Pintar A., (2010). UV- and visible-light assisted photocatalytic oxidation of a pesticide over TiO₂ catalysts modified with neodymium and nitrogen. *In CD ROM of Full Texts of 7th European Congress of Chemical Engineering ECCE and 19th International Congress of Chemical and Process Engineering CHISA 2010, 28 August – 1 September 2010, Prague, Czech Republic.*

Lukač P., Földesová M. & Svetlík Š., (2005). Synergistic effect of chemical and thermical treatment on the structure and sorption properties of natural and chemically modified Slovak zeolite. *Petroleum & Coal* 47, (1) pp. 17-21.

Mujeriego R & Asano, T., (1999). The role of advanced treatment in wastewater reclamation and reuse. *Water Science and Technology* 40, (4-5) pp. 1-9.

Müller J. A., (2000). Pretreatment processes for the recycling and reuse of sewage sludge. *Water Sci. Technol.* 42, (9) pp. 167-174.

Munter R., (2001). Advanced oxidation processes - current status and prospects. Proc. Estonian Acad. *Sci. Chem.* 50, (2) pp. 59-80.

Oller I., Sirtori C., Klamerth N. & Zapata A., (2009). Decontamination of industrial wastewater by advanced oxidation processes coupled with biotreatment. *In. Proceedings from INNOVA-MED Conference. Innovative processes and practices for wastewater treatment and re-use in the Mediterranean region. 8.-9 October 2009, Chirona, Spain.* pp. 57-60.

Park K. Y., Ahn K. H., Maeng S. K., Hwang J. H. & Kwon J. H., (2003). Feasibility of sludge ozonation for stabilisation and conditioning. *Ozone Sci. Eng.* 25 (1) pp. 73–80.

Poyotos J. M., Muñio M. M., Almecija M. C., Torres J. C., Hontoria E. & Osorio F., (2010). Advanced oxidation processes for wastewater treatment: state of the art. *Water air soil pollut* 205, pp. 187-204.

Rice R. G., (1997). Applications of ozone for industrial wastewater treatment – A review. *Ozone Science and Engineering.* 18, pp. 477-515.

Rodrígeus S. M., (2009). Advanced technologies for wastewater treatment. *In. Proceedings from INNOVA-MED Conference. Innovative processes and practices for wastewater treatment and re-use in the Mediterranean region. 8.-9 October 2009, Chirona, Spain.* pp. 18-22.

Rulkens W., (2008). Increasing significance of advanced physical/chemical processes in the development and application of sustainable wastewater treatment systems. *Frontiers of Environmental Science and Engineering in China* 2 (4) pp. 385-396.

Spanjer H., Vanrolleghem P. A., Olson G. & Dold P.L., (2000). Respirometry in Control of the Activated Sludge Process: Principles. *IAWQ Scientific and Technical Report* No. 7. J. W. Arrowsmith Ltd, Bristol, England.

Tomlinson T. G., Boon A. G., & Trotman C. N. A., (1966) Inhibition of nitrification in the activated sludge process of sewage disposal. *J. Appl. Bacteriol.* 29 (2) pp. 266-291.

Valdés H. & Zahor C. A., (2006). Ozonation of benzothiazole saturated-activated carbons: Influence of carbon chemical surface properties. *J. Hazard Mater.* B153, pp. 1042-1048.

Valdes H., Farfán, V. J., Manoli, J. A. & Zaror C. A., (2009). Catalytic ozone aqueous decomposition promoted by natural zeolite and volcanic sand. *J. Hazard Mater.* 165, pp. 915-922.

Valdés H., Murillo F. A., Manoli J.A. & Zahor C. A., (2008). Heterogeneous catalytic ozonation of benzothiazole aqueous solution promoted by volcanic sand. *J. Hazard Mater.* 153, pp. 1036-1042.

6

Modelling of Chemical Alteration of Cement Materials in Radioactive Waste Repository Environment

Daisuke Sugiyama
Nuclear Technology Research Laboratory,
Central Research Institute of Electric Power Industry
Japan

1. Introduction

Cement is a potential waste packaging, backfilling and constructing material for the disposal of radioactive waste. The physical properties of cement materials such as low permeability and low diffusivity in their matrices reduce the migration of radionuclides from a cementitious repository. Also, under a high-pH condition provided by the leaching of the components of cement hydrates, the solubility is low and the sorption distribution ratio is high for many radionuclides, so that the release of radionuclides from radioactive waste is restricted. Therefore, cement materials are expected to enable both the physical and chemical containments of long-term radioactive waste in disposal systems (TRU Coordination Office, 2000).

Under geological conditions, cement materials alter due to various reactions such as dissolution into groundwater and secondary mineral formation caused by chemical components in groundwater. The containment properties of cement materials are affected by these reactions. Also, the leached high-pH solution with alkaline components from the cement materials affects the physical and chemical properties of bentonite and the surrounding rocks. Therefore, for the long-term safety assessment of radioactive waste disposal, it is necessary to develop a methodology to estimate the long-term evolution of the cementitious repository system. The chemistry of the $CaO\text{-}SiO_2\text{-}H_2O$ (C-S-H) system is a key parameter since it is suggested to be responsible for the high-pH condition in cements and is important in discussing the high-pH chemical condition in the long-term assessment of a repository environment (Atkinson et al., 1985; Atkinson, 1985). The author therefore has been developing a series of predictive calculation models based on a discussion of the incongruent dissolution/precipitation of the C-S-H system (Sugiyama & Fujita, 2006; Sugiyama et al., 2007; Sugiyama, 2008).

In this study, the alteration of cement materials in an underground repository environment is discussed. Cement materials come in contact with groundwater and some secondary minerals are expected to precipitate in the repository environment. The precipitation of calcite and its effects on the alteration of cement materials should be key issues in assessing the long-term performance of cement materials. There have been some experimental studies on these issues, in which the precipitation of calcite from calcium bicarbonate solution passing through cracks in concrete and the leaching behaviour of components from cementitious materials were observed (Harris et al., 1998; Glasser et al., 2001). It was observed that calcite is mostly precipitated on the surface of the cracks, filling the cracks, and it was suggested that the thin layers of low-porosity calcite produced act as a diffusion barrier limiting contact between the cement and the solution (Harris et al., 1998; Glasser et al., 2001). Brodersen (2003) simulated the obtained experimental results using the CRACK2 computation model. Also, Lagneau and van der Lee (2005) simulated the clogging effect due to the precipitation of secondary minerals using the HYTEC code and showed that the precipitation leads to a reduction in porosity, which reduces the diffusive migration rate of chemical species. Burnol et al. (2005) discussed the precipitation behaviour of some minerals at a concrete/clay interface using various reactive transport codes. Marty et al. (2009) investigated the modelling of concrete/clay interactions under a geological disposal condition and discussed an occlusion due to the secondary mineral precipitation including calcite. However, the incongruent dissolution of the calcium-silicate hydrate (C-S-H) phase, which is one of the principle components of cementitious materials, was not sufficiently discussed in the previous modelling studies. Brodersen (2003) disregarded the leaching of silicic acid, and in the other studies, only three C-S-H phases (C-S-H 1.8, C-S-H 1.1 and C-S-H 0.8 (Lagneau & van der Lee, 2005; Burnol et al., 2005), and C-S-H 1.6, C-S-H 1.2 and C-S-H 0.8 (Marty et al., 2009) were included discretely in their modelling.

In this study, a reactive transport computational code, in which a geochemical model including the thermodynamic incongruent dissolution model of C-S-H (Sugiyama & Fujita, 2006) is coupled with the advection-diffusion/dispersion equation including the evolution of the hydraulic properties of the solid cement matrix due to the leaching and precipitation of components, has been developed. A series of experiments on the alteration of hydrated ordinary portland cement (OPC) and low-heat portland cement containing 30 wt% fly ash (FAC) monoliths in deionised water and sodium bicarbonate (NaHCO$_3$) solution were carried out and the model was optimised on the basis of the observations. Using the developed code, some implications for the long-term performance assessment of radioactive waste disposal systems were discussed.

2. Development of calculation code CCT-P

2.1 Model description

A coupling transport and chemical equilibrium calculation code, coupled chemical equilibria-mass transport code for porous media (CCT-P), is developed to predict the alteration behaviour of cement materials. The calculations using CCT-P are performed with a two-step procedure involving a nonreactive transport step (using Eqs. (3)-(8)) followed by a chemical equilibrium calculation. All thermodynamic modelling calculations were carried out using the thermodynamic database HATCHES (Bond et al., 1997) Ver. NEA15.

The thermodynamic incongruent dissolution model of C-S-H, proposed by Sugiyama & Fujita (2006), is employed in the modelling approach. In this model, C-S-H is assumed to be a binary nonideal solid solution of Ca(OH)$_2$ and SiO$_2$, and the log K values of the model end members of the solid solution are given as functions of the Ca/Si ratio of C-S-H (Sugiyama & Fujita, 2006):

$$\log K_s = \frac{1}{1+y} \cdot \log K_{s0} - \frac{1}{1+y} \cdot \log\frac{1}{1+y} + \frac{y}{(1+y)^2} \cdot \left[A'_{s0} + A'_{s1}\left(\frac{1-y}{1+y}\right) + A'_{s2}\left(\frac{1-y}{1+y}\right)^2 \right], \qquad (1)$$

$$\log K_c = \frac{y}{1+y} \cdot \log K_{c0} - \frac{y}{1+y} \cdot \log\frac{y}{1+y} + \frac{y}{(1+y)^2} \cdot \left[A'_{c0} + A'_{c1}\left(\frac{1-y}{1+y}\right) + A'_{c2}\left(\frac{1-y}{1+y}\right)^2 \right], \qquad (2)$$

y = Ca/Si ratio in C-S-H, $\log K_{s0} = -2.710$, $\log K_{c0} = 22.81$ (Bond et al., 1997).

At Ca/Si ≤ 0.461, $\log K_s = \log K_{s0} - \log(1+y)$.

At $1.755 \leq$ Ca/Si, $\log K_s = -7.853$, $\log K_c = 22.81$.

The fitted empirical parameters (A_{ij}) are shown in Table 1. This model can predict the equilibria of the incongruent dissolution and precipitation with a continuous change in the Ca/Si ratio of the solid phase by a series of calculations, in which the quantities of the dissolved/precipitated end members are calculated stepwise, so that the quantities and compositions of the solid and liquid phases and the conditional solubility constants used in the next step can be estimated (Sugiyama & Fujita, 2006). CCT-P contains the geochemical code PHREEQE (Parkhurst et al., 1980) to calculate the chemical equilibrium, and the C-S-H model is employed to calculate the incongruent dissolution and precipitation by iterative calculations, because the simplicity of its numerical description allows its inclusion in chemical equilibrium calculations based on the common approach of using the law of mass action ($\log K$) in PHREEQE (Parkhurst et al., 1980).

Fig. 1 (a) shows a schematic representation of the transport model used in CCT-P. To calculate the mass transport in the porous media, a one-dimensional advection /dispersion/diffusion equation is employed in CCT-P:

$$\frac{\partial}{\partial x}\{D_e(x,t)\cdot\frac{\partial C(x,t)}{\partial x}\} - V_d \cdot \frac{\partial C(x,t)}{\partial x} = \frac{\partial\{\phi(x,t)\cdot R_d(x,t)\cdot C(x,t)\}}{\partial t} - S_{eq}(x,t), \qquad (3)$$

where
C: concentration of aqueous species,
t : time,
ϕ : porosity,
V_d: velocity of flow in matrix,
D_e: effective diffusion coefficient in matrix,
S_{eq} : source term given by chemical equilibrium calculation within matrix,

R_d: retardation factor ($R_d(t) = 1 + \rho \cdot K_d \cdot \left(\frac{1-\phi(t)}{\phi(t)}\right)$),

ρ : density,
K_d : distribution coefficient.

End member	SiO$_2$			Ca(OH)$_2$		
A_{ij}	A_{s0}	A_{s1}	A_{s2}	A_{c0}	A_{c1}	A_{c2}
Ca/Si ≤ 0.833	-18.623	57.754	-58.241	37.019	-36.724	164.17
Ca/Si > 0.833	-18.656	49.712	25.033	36.937	-7.8302	-50.792

Table 1. Fitted values of empirical parameters (Sugiyama & Fujita, 2006)

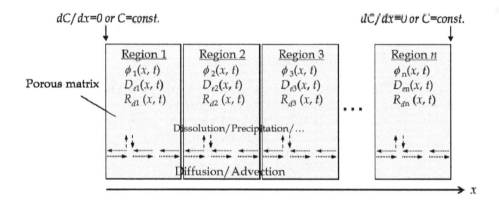

(a) Schematic representation of transport model

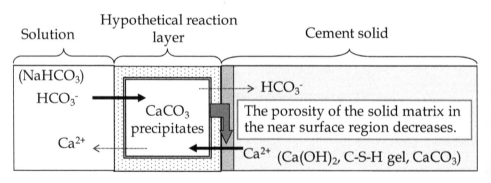

(b) Schematic representation of hypothetical reaction layer model

Fig. 1. Transport model in CCT-P

The evolution of the hydraulic properties of the solid cement matrix due to the leaching and precipitation of components is also considered in the calculation code. The component minerals in the solid are leached from its surface into the solution, and are precipitated on the surface and in the matrix of the solid. In this study, it is considered that the porosity of the solid matrix increases or decreases as the component minerals are dissolved and leached or precipitated, respectively. This is described by the following equations:

$$\phi(t) = 1 - (1 - \phi(0)) \cdot \frac{V_{solid}(t)}{V_{solid}(0)} , \tag{4}$$

$$V_{solid}(t) = \sum_{i:solid} CS_i(t) \cdot v_{mol_i} + V_{solid,static} , \tag{5}$$

where
V_{solid}: volume of solid phase,
CS: molarity of component mineral,
v_{mol}: molar volume of component mineral ($v_{mol\ Ca(OH)2}$ = 0.0331 dm³ mol⁻¹,
$v_{mol\ SiO2}$ = 0.0273 dm³ mol⁻¹, $v_{mol\ CaCO3}$ = 0.0369 dm³ mol⁻¹),
$V_{solid,static}$: volume of insoluble residual solid phase.

In CCT-P, the diffusion coefficient in the altered region of the solid matrix can be described as a function of porosity using formulas based on the study by Haga et al. (2005):

$$D(t) = D(0) \cdot \left(\frac{\phi(t)}{\phi(0)} \right)^n . \tag{6}$$

In this study, n = 2 (Haga et al., 2005) is assumed when calculating the evolution of the diffusion coefficient.

At the boundaries of the regions (Fig. 1 (a)), the advection/dispersion/diffusion equations (Eq. (3)) in adjacent regions are connected as follows:

$$J_{inner} = J_{outer} \tag{7}$$

where

J: flux of aqueous species,

$$J_{inner} = -D_e(x,t)_{inner} \cdot \left. \frac{\partial C(x,t)}{\partial x} \right|_{inner} + V_{d_{inner}} \cdot C(x,t)_{boundary} ,$$

$$J_{outer} = -D_e(x,t)_{outer} \cdot \left. \frac{\partial C(x,t)}{\partial x} \right|_{outer} + V_{d_{outer}} \cdot C(x,t)_{boundary} .$$

The subscripts *inner* and *outer* denote the values in the adjacent inner and outer cells, respectively.

In this study, the effects of the precipitation of calcite on the alteration of cement material are discussed. In the experiment, it was observed that calcite was mostly precipitated on the surface of the cement monolith, and the thin layers of low-porosity calcite produced acted as a diffusion barrier limiting contact between the solid cement and the solution (see subsection 4.1). Calcite precipitated in the pore solution of the cement solid matrix and also in the bulk solution near the interface between the cement solid and the solution. To describe this, a hypothetical reaction layer model is proposed and included in CCT-P, as shown in Fig. 1 (b). In this model, a less-soluble or insoluble secondary phase is formed in a hypothetical reaction layer and precipitates in the vicinity of the surface of a solid phase. A hypothetical reaction layer is added between the solid and solution phases (outside of the solid phase interested), and calcium ions leached from the solid cement react with bicarbonate ions in the hypothetical reaction layer to form calcite. The formed calcite precipitates on the surface of the solid cement and the porosity of the solid matrix in the near-surface region decreases, then the precipitated layer of low-porosity restricts the diffusion mass transfer. This is described by the following equation:

$$ -D_e(x,t)_{cement} \cdot \frac{\partial C(x,t)}{\partial x}\bigg|_{cement} = \left(\frac{\phi(t)}{\phi(0)}\right)_{surface} \cdot \left\{ -D_e(x,t)_{HRL} \cdot \frac{\partial C(x,t)}{\partial x}\bigg|_{HRL} \right\}, \tag{8}$$

where the subscripts *cement*, *surface* and *HRL* denote the bulk cement region, the near-surface layer of the cement region in which the porosity decreases and the hypothetical reaction layer, respectively. $\phi(t)_{surface}$ is calculated using Eq. (4), assuming that all secondary precipitates formed in the HRL precipitate on the near-surface layer with a thickness of L_{shell} and contribute to adjust porosity. The thickness of the near-surface layer (L_{shell}) is provided as a parameter in CCT-P. Note that the code refers to $\phi(t)_{surface}$ only in calculations using Eq. (8) connecting regions at the boundary. The term $(\phi(t)/\phi(0))_{surface}$ is used to describe that contact between the solid cement and the solution is limited by the calcite precipitate, which acts as a barrier.

2.2 Numerical method

The one-dimensional advection/dispersion/diffusion equation (Eq. (3)) employed in CCT-P is numerically solved by an implicit finite-difference method. Eq. (3) can be described in the generic form

$$\frac{\partial C_i}{\partial t} = A\frac{\partial^2 C_i}{\partial x^2} - B\frac{\partial C_i}{\partial x} - CC \cdot C_i + D, \tag{9}$$

where

$$A = \frac{D_e(x,t)}{\phi(x,t)\cdot R_d(x,t)},$$

$$B = \frac{1}{\phi(x,t)\cdot R_d(x,t)}\{-\frac{\partial D_e(x,t)}{\partial x}+V_d\},$$

$$CC = \frac{1}{\phi(x,t)\cdot R_d(x,t)}\cdot\frac{\partial\{\phi(x,t)\cdot R_d(x,t)\}}{\partial t},$$

$$D = \frac{S_{eq}(x,t)}{\phi(x,t)\cdot R_d(x,t)}.$$

The finite-difference equation for Eq. (9) is given as

$$ -a_j C_{j-1}^n + b_j C_j^n - c_j C_{j+1}^n = d_j, \tag{10}$$

$$d_j = a_j C_{j-1}^o + b_j^* C_j^o + c_j C_{j+1}^o + D_j \cdot (\Delta x_j + \Delta x_{j+1}), \tag{11}$$

$j = 2, \ldots$ N-1. (N: total number of meshes)

a_j, b_j, c_j and b_j^* are given as

$$a_j = \frac{A_j}{\Delta x_j} - \frac{1}{2} B_j$$

$$b_j = A_j \cdot \left(\frac{1}{\Delta x_j} + \frac{1}{\Delta x_{j+1}} \right) + \frac{1}{2} CC_j \cdot (\Delta x_j + \Delta x_{j+1}) + \frac{1}{\Delta t} (\Delta x_j + \Delta x_{j+1})$$

$$c_j = \frac{A_j}{\Delta x_{j+1}} - \frac{1}{2} B_j$$

$$b_j^* = - A_j \cdot \left(\frac{1}{\Delta x_j} + \frac{1}{\Delta x_{j+1}} \right) - \frac{1}{2} CC_j \cdot (\Delta x_j + \Delta x_{j+1}) + \frac{1}{\Delta t} (\Delta x_j + \Delta x_{j+1})$$

Spatial discretization is performed by centered-in-space differencing and temporal discretization is performed by Crank-Nicholson (centered-in-time) differencing. A, B, CC and D in Eq. (9) are also described as the differenced forms

$$A_j = \frac{D_{ej}}{\phi_j \cdot R_{dj}},$$

$$B_j = \frac{1}{\phi_j \cdot R_{dj}} \left\{ -\frac{D_{ej+1} - D_{ej}}{\Delta x_j} + V_d \right\},$$

$$CC_j = \frac{1}{\phi_j \cdot R_{dj}} \cdot \frac{\phi_j^n \cdot R_{dj}^n - \phi_j^o \cdot R_{dj}^o}{\Delta t},$$

$$D_j = \frac{S_{eqj}}{\phi_j \cdot R_{dj}}.$$

(The superscripts 'n' and 'o' denote the newest value and an old value obtained at t-Δt, respectively. Porosity, effective diffusion coefficient and retardation factor without superscript are the newest values.)

For the boundary condition, the finite-difference equations are given as follows.

Upper boundary (Dirichlet boundary condition)

$$b_1 C_1^n - c_1 C_2^n = d_1, \tag{12}$$

$$d_1 = a_1 (C_b^n + C_b^o) + b_1^* C_1^o + c_1 C_2^o + D_1 \cdot (\Delta x_1 + \Delta x_2). \tag{13}$$

Upper boundary (Flux boundary condition)

$$(b_1 - a_1 \frac{\frac{D_{e,i}}{\Delta x_1}}{v_{db} + \frac{D_{e,i}}{\Delta x_1}}) C_1^n - c_1 C_2^n = d_1 , \tag{14}$$

$$d_1 = a_1 \frac{J_b^n + J_b^o}{v_{db} + \frac{D_{e,i}}{\Delta x_1}} + \left[\frac{a_1 \frac{D_{e,i}}{\Delta x_1}}{v_{db} + \frac{D_{e,i}}{\Delta x_1}} + b_1^* \right] C_1^o + a_1 C_2^o + D_1 \cdot (\Delta x_1 + \Delta x_2) . \tag{15}$$

Lower boundary (Dirichlet boundary condition)

$$-a_{N-1} C_{N-2}^n + b_{N-1} C_{N-1}^n = d_{N-1} , \tag{16}$$

$$d_{N-1} = c_{N-1} (C_b^n + C_b^o) + a_{N-1} C_{N-2}^n + b_{N-1}^* C_{N-1}^o + D_{N-1} \cdot (\Delta x_{N-1} + \Delta x_N) . \tag{17}$$

Lower boundary (Neumann boundary condition)

$$-a_{N-1} C_{N-2}^n + (b_{N-1} - c_{N-1}) C_{N-1}^n = d_{N-1} , \tag{18}$$

$$d_{N-1} = a_{N-1} C_{N-2}^o + (b_{N-1}^* + c_{N-1}) C_{N-1}^o + D_{N-1} \cdot (\Delta x_{N-1} + \Delta x_N) . \tag{19}$$

At the boundaries of the regions, the advection/dispersion/diffusion equations in adjacent regions are connected using the following finite-diffusion equation:

$$-\frac{D_{e,j-1}}{\Delta x_{j-1}} C_{j-1}^n + (\frac{D_{e,j-1}}{\Delta x_j} + \frac{D_{e,j}}{\Delta x_{j+1}} - v_{dj,inner} + v_{dj,outer}) C_j^n - \frac{D_{e,j}}{\Delta x_{j+1}} C_{j+1}^n$$

$$= \frac{D_{e,j-1}}{\Delta x_j} C_{j-1}^o - (\frac{D_{e,j-1}}{\Delta x_j} + \frac{D_{e,j}}{\Delta x_{j+1}} - v_{dj,inner} + v_{dj,outer}) C_j^o - \frac{D_{e,j}}{\Delta x_{j+1}} C_{j+1}^o . \tag{20}$$

In CCT-P, these equations are solved by the backward substitution method.

3. Experimental

3.1 Materials

A series of experiments on the alteration of hydrated cement monoliths in deionised water and sodium bicarbonate solution were carried out in this study. Solid monolith samples of ordinary portland cement (OPC) and low-heat portland cement containing 30 wt% fly ash (FAC) were prepared for use in a so-called tank leaching experiment.

The chemical compositions of OPC and FAC used in this study are shown in Table 2. The cement was hydrated at a water/cement clinker mixing ratio of 0.35. The hydrated materials were then cured in water at 50 °C to enhance hydration (Taylor, 1997) and reduce the effects of unhydrated phases for 91 days.

The hydrated solid samples were analysed by X-ray diffraction (XRD). The results are shown in Table 3, and the predominant phases were portlandite, C-S-H gel and ettringite for all solid cement hydrate samples. In FAC hydrate, XRD identified katoite ($Ca_3Al_2(SiO_4)(OH)_8$) as a minor mineral and quartz (SiO_2) as an unhydrated phase. The amount of $Ca(OH)_2$ in the hydrated solid samples was quantified by DTA and is shown in Table 3. The porosity of the solid matrix of cement samples and the pore size distribution were measured by mercury intrusion porosimetry, and these data are shown in Table 3 and Fig. 2.

3.2 Tank leaching experiment

Hardened OPC and FAC after curing were cut into $20 \times 20 \times 10$ mm blocks using a diamond cutter. The cement monolith samples were set in the acrylic cells shown in Fig. 3 with a circular window of 12 mm diameter and placed in contact with the solution. The exposure of only one of the faces of each monolith to the aqueous solution in the alteration (leaching/precipitation) experiments simplifies the subsequent analysis of the distribution of components in the cement monolith. Each cell containing a cement monolith was placed in a vessel to which 80 cm^3 of deionised water or $NaHCO_3$ solution at room temperature was added. The initial concentrations of $NaHCO_3$ were 6×10^{-5}, 1×10^{-4}, 6×10^{-4}, 1×10^{-3}, and 6×10^{-3} mol dm^{-3}. The solution and solid samples were separated and the monolith samples were again placed in contact with fresh deionised water or $NaHCO_3$ solution after 1 week and then every 4-5 weeks. All experiments were prepared in nitrogen-filled glovebox in triplicates.

The concentrations of calcium and silica in each separated solution were measured by inductively coupled plasma atomic emission spectroscopy (ICP-AES) after passing it through a 0.45 μm membrane filter. Each filtered solution was acidified and diluted with known volumes of nitric acid (HNO_3) solution and distilled water before ICP-AES analysis to ensure that the concentrations of the target and matrix elements were appropriate for the analysis. The carbonate ion concentration in the solution was measured by ion chromatography. The surface of the monolith sample after 76 weeks was observed by scanning electron microscopy (SEM) and the cross section of the solid was analysed by energy-dispersive X-ray analysis (EDX analysis) to observe the distribution of components in a cement matrix.

Oxide composition / wt%	SiO_2	Al_2O_3	Fe_2O_3	CaO	MgO	SO_3	Na_2O	K_2O
OPC	21.4	5.4	2.7	64.8	1.5	2.1	0.3	0.5
FAC	32.9	9.5	3.8	46.9	1.2	1.9	0.8	0.6

Table 2. Oxide composition of cement clinker used in experiment

Sample	Minerals (Identified by XRD)	$Ca(OH)_2$ / wt% (Measured by DTA)	Porosity / % (Measured by mercury intrusion porosimetry)
OPC	$Ca(OH)_2$, C-S-H gel, Ettringite	17.4	12.0
FAC	$Ca(OH)_2$, C-S-H gel, Ettringite, Quarts, Katoite	4.0	28.9

Table 3. Analytical results of XRD, DTA and mercury intrusion porosimetry of hydrated cement sample

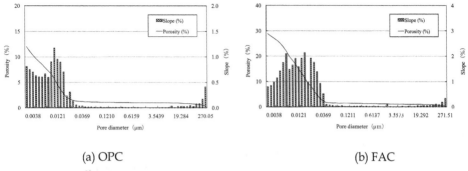

(a) OPC   (b) FAC

Fig. 2. Pore size distributions of cement monolith samples

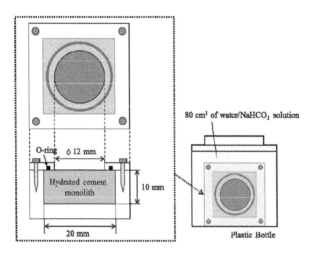

Fig. 3. Experimental cell for cement alteration experiment

4. Results

4.1 Cement monolith alteration experiments

The amount of calcium leached from the OPC monolith sample was calculated using the measured aqueous composition and is shown in Fig. 4 (a) as a function of time. In deionised water, calcium was leached from the surface of the OPC solid monolith and the leaching rate decreased slightly with time. This result suggests that in the early stage of alteration, the dissolution equilibrium of portlandite ($Ca(OH)_2$) dominated the leaching of calcium, and in the late stage, the incongruent dissolution of C-S-H gel in the altered surface region dominated the calcium leaching. In sodium bicarbonate solution at a $NaHCO_3$ concentration higher than 6×10^{-4} mol dm^{-3}, a reduction in the rate of calcium leaching was observed. As the concentration of $NaHCO_3$ increased, the rate of calcium leaching decreased and was restricted significantly in the later stage of the experiment. The leaching of calcium was always inhibited at the $NaHCO_3$ concentration of 6×10^{-3} mol dm^{-3} during the experiment.

In the FAC experiments shown in Fig. 4 (b), the trend of calcium leaching was observed to be similar to that in OPC experiments; however, the rate of calcium leaching was lower in the cases of FAC than in the cases of OPC. FAC inhibited the leaching of calcium in 6×10^{-4} mol dm^{-3} NaHCO$_3$ solution earlier than OPC. The leaching of calcium was inhibited from the early stage of the experiment at the NaHCO$_3$ concentration of $> 1 \times 10^{-3}$ mol dm^{-3}.

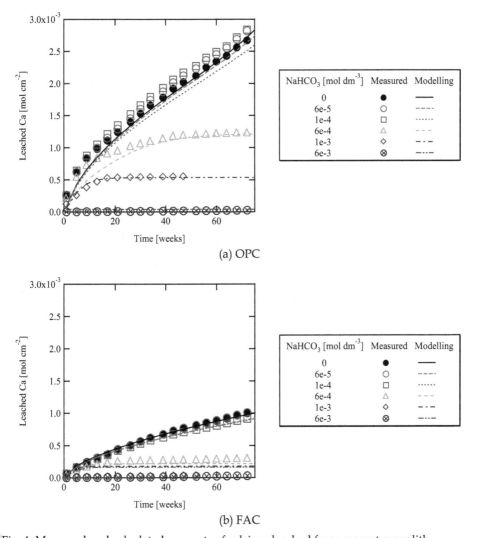

(a) OPC

(b) FAC

Fig. 4. Measured and calculated amounts of calcium leached from cement monolith

For all the cases of alteration in NaHCO$_3$ solution, a secondary crystalline precipitate was formed on the surface of the solids, as shown in Fig. 5. The precipitate was identified as calcite by XRD analysis. A denser calcite layer was formed at higher NaHCO$_3$ concentrations of 1×10^{-3} and 6×10^{-3} mol dm^{-3}.

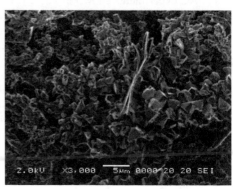

(a) OPC (in distilled water)

(b) OPC (in 1×10^{-4} mol dm^{-3} NaHCO$_3$)

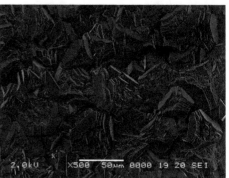

(c) OPC (in 6×10^{-4} mol dm^{-3} NaHCO$_3$)

(d) OPC (in 6×10^{-4} mol dm^{-3} NaHCO$_3$)
(Filmy layer of calcite precipitate)

(e) OPC (in 6×10^{-3} mol dm^{-3} NaHCO$_3$)

(f) FAC (in distilled water)

(g) FAC (in 1×10^{-4} mol dm^{-3} NaHCO$_3$)

(h) FAC (in 6×10^{-4} mol dm^{-3} NaHCO$_3$)

(i) FAC (in 6×10^{-4} mol dm^{-3} NaHCO$_3$)
(Filmy layer of calcite precipitate)

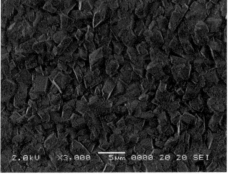

(j) FAC (in 6×10^{-3} mol dm^{-3} NaHCO$_3$)

Fig. 5. SEM observation on surface of cement monolith

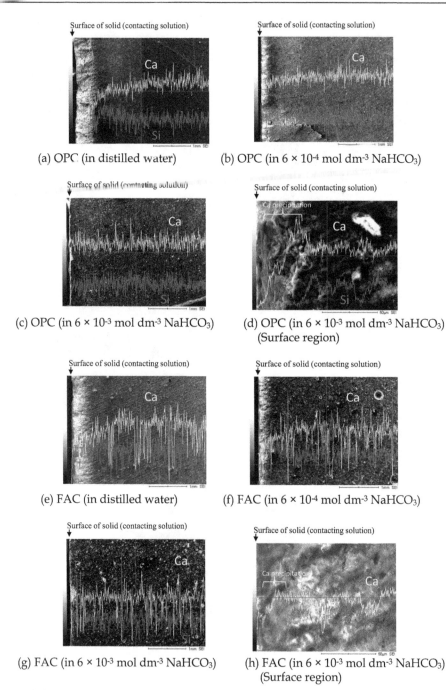

Fig. 6. Calcium concentration profiles at cross section of altered cement sample after 76 weeks obtained by EDX analysis

The calcium concentration at the cross section of the divided altered solid samples analysed by EDX analysis is shown in Fig. 6. In deionised water, calcium was leached from the surface and the Ca/Si ratio of the solid decreased in the surface region up to a depth of around 2 mm. In sodium bicarbonate solution at a higher $NaHCO_3$ concentration of 6×10^{-4} mol dm^{-3}, the region with decreasing Ca/Si ratio of the solid was small up to a depth of around 1.0 mm. In contrast, it was observed that a small amount of calcium leached in 6×10^{-3} mol dm^{-3} sodium bicarbonate solutions. In 6×10^{-3} mol dm^{-3} $NaHCO_3$ solutions, a layer of accumulated calcium was formed at the vicinity of the surface of the solid, as shown in Figs. 6 (d) and (h). It is suggested that calcite precipitated on the surface of the monoliths in sodium bicarbonate solution and the secondary calcite precipitation concentrated in the near-surface layer. Thus, the secondary calcite precipitation restricted the leaching of calcium from the solid by limiting the contact between the solid cement and the solution.

The amount of calcite precipitated on the cement monolith sample was calculated using the measured carbonate ion concentration in the separated solution and is shown in Fig. 7 as a function of time. In $< 6 \times 10^{-4}$ mol dm^{-3} $NaHCO_3$ solutions, more calcite precipitated at higher $NaHCO_3$ concentrations. In the cases where a restriction of calcium leaching in $> 6 \times 10^{-4}$ mol dm^{-3} $NaHCO_3$ solutions was observed, the calcite precipitation rate decreased.

4.2 Modelling calculation of cement alteration experiment

The cement alteration experiments were simulated using the developed model and the CCT-P calculation code. First, the mineral compositions of the hydrated cements were derived by an approach based on that described by Glasser et al. (1987) with some modifications. Glasser et al. calculated the equilibrium phase distribution by considering the phases of $Ca(OH)_2$, C-S-H, hydrotalcite and monosulphate (Glasser et al., 1987). The hydrated OPC and FAC in the experiments were dominantly composed of portlandite, C-S-H gel, ettringite and a small amount of other minerals, and the amount of portlandite was measured by DTA. Thus, in this study, the mineral compositions of the hydrated OPC and FAC were calculated by the following approach.

For OPC:

Step 1. The mineral assembly in the hydrated OPC was described with portlandite $(Ca(OH)_2)$, C-S-H, ettringite, brucite $(Mg(OH)_2)$, NaOH and KOH. Na_2O, K_2O and MgO were assumed to be completely hydrated to NaOH, KOH and brucite, respectively.

Step 2. The amount of portlandite was estimated by DTA of the hydrated OPC sample used in the experiment in this study.

Step 3. SiO_2 was assumed to be taken up by the C-S-H gel with Ca/Si = 1.755 (Sugiyama & Fujita, 2006), and C-S-H was described using the model proposed by Sugiyama and Fujita (2006).

Step 4. The remaining CaO was assumed to be taken up by ettringite. The remaining Al_2O_3 was assumed to be amorphous alumina gel or taken up by some phases (e.g., C-A-S-H gel), although no excess Al was included in the following modelling calculations.

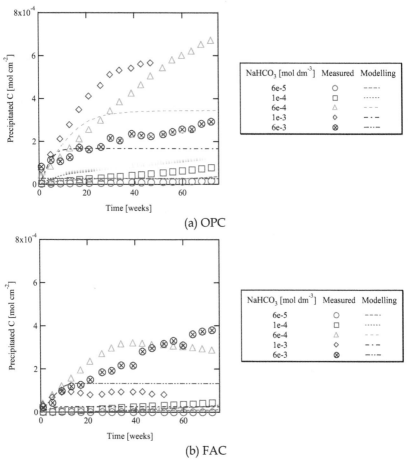

(a) OPC

(b) FAC

Fig. 7. Measured and calculated amounts of calcite precipitated on cement monolith

For FAC:

Step 1. The mineral assembly in the hydrated FAC was described with portlandite ($Ca(OH)_2$), C-S-H, ettringite, brucite, katoite, NaOH, KOH and quartz. Na_2O, K_2O and MgO were assumed to be completely hydrated to NaOH, KOH and brucite, respectively.

Step 2. SO_3 was assumed to be taken up by ettringite. The remaining Al_2O_3 was assumed to be taken up by katoite.

Step 3. The remaining CaO was assumed to be taken up by portlandite and C-S-H gel with Ca/Si = 1.755 (Sugiyama & Fujita, 2006). The amount of portlandite was estimated by DTA of the hydrated FAC sample used in the experiment in this study.

Step 4. SiO_2 was assumed to be taken up by the C-S-H gel with Ca/Si = 1.755 (Sugiyama & Fujita, 2006).

Step 5. The remaining SiO_2 was assumed to be quartz, which is an unhydrated component in fly ash.

The mineral compositions of the OPC and FAC hydrates calculated by the method above are shown in Table 4. Then, the cement hydrates are simply described as the assembly of $Ca(OH)_2$ and C-S-H gel in this modelling study to reduce the load of calculation.

The calculation system was defined using the parameters in Table 5 and the model in Fig. 8. A constant boundary condition was imposed at the surface of the solid, which described the experimental condition of the replacement of the solution, and at the other end, a closed boundary condition was prescribed. Diffusion transport was calculated and advection transport was not considered in this simulation. As described in subsection 2.1, the HRL was added at the boundary between the solid and the solution to simulate the calcite precipitation on the surface of the monolith in the model. The amount of calcium leached from the solid into the solution was calculated by integrating the flux of calcium at the surface of the solid ($x = 0$).

[mol kg^{-1}]

Cement	Quartz	Katoite	Brucite	Ettringite	Ca(OH)$_2$	C-S-H gel (Ca/Si = 1.755)	NaOH	KOH
OPC	-	-	0.28	0.29	2.35	2.76	0.07	0.08
FAC	1.65	0.72	0.25	0.07	0.54	2.27	0.21	0.10

Table 4. Calculated mineral compositions of cement hydrates

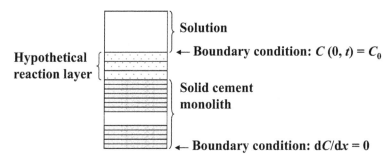

Fig. 8. Analysis model used for OPC alteration experiment

Cement		OPC	FAC
Hypothetical reaction layer (HRL)	Thickness of region [mm]	0.3	0.3
	Thickness of each grid layer [mm]	0.1	0.1
	Initial D_e [m^2 s^{-1}]	$8.0 \times 10^{-10*}$	$8.0 \times 10^{-10*}$
Cement hydrate solid	Thickness of region [mm]	10	10
	Thickness of each grid layer [mm]	0.267	0.200
	Initial D_e [m^2 s^{-1}]	$1.9 \times 10^{-11**}$	$7.4 \times 10^{-12**}$
	Initial porosity	0.120	0.289

* An infinite dilution diffusion coefficient for calcium was assumed in the calculation.
** The initial effective diffusion coefficient in the cement solid matrix was provided by fitting calculation using the leaching data of calcium from the cement monolith in distilled water.

Table 5. Calculation parameters used for modelling OPC alteration experiment

The spatial resolution and the initial effective diffusion coefficient were optimised by preliminary sensitivity calculations to reproduce the rate of calcium leaching in the experiments in distilled water. The fitted initial effective diffusion coefficients given in Table 5 are comparable with the values measured by Haga et al. (2005) and Yasuda et al. (2002). The time step size was given in consideration of a Neumann criterion (Marty et al., 2009; Sousa, 2003; Hindmarsh & Gresho, 1984):

$$\frac{2D_p \cdot \Delta t}{\Delta x^2} \leq 1, \tag{21}$$

where D_p ($= D_e/\phi$) is the pore diffusion coefficient, and Δx and Δt refer to the mesh size in the solid region and the time step.

For modelling using the two-step procedure, it is suggested that the choice of the space discretization affects the calculation results (Marty et al., 2009). In this study, the sensitivity of calculation to the mesh size was analysed. The sensitivity calculation results are shown in Fig. 9. In the case of OPC (Fig. 9 (a)), coarser mesh sizes gave a lower rate of calcium leaching and finer mesh sizes degraded the stability of numerical solutions. The optimised mesh size in this case is 0.267 mm, which reproduced the measured result using a diffusion coefficient (1.9×10^{-11} m² s⁻¹) compatible with the experimentally determined value (Haga et al., 2005). In the case of FAC (Fig. 9 (b)), the calculation result was less sensitive to the space discretization, though a behaviour similar to that of OPC was observed.

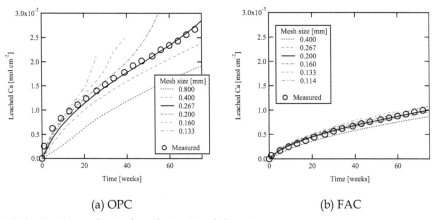

(a) OPC   (b) FAC

Fig. 9. Sensitivity analysis of mesh size in solid region

The thickness of the HRL was optimised by sensitivity calculations given in Fig. 10. It is considered that a smaller HRL gives lesser disturbance to the calculation. The optimised HRL size was chosen to be 0.3 mm since too fine HRL could affect the stability of the numerical solution. It is noted that FAC calculation was insensitive to the HRL thickness (Fig. 10 (b)).

To model the alteration in the NaHCO₃ solutions, the thickness of the near-surface layer (L_{shell}) must be determined. Parametric analysis was carried in this study and is discussed in subsection 5.1. In this subsection, the best-fit results are described with the optimised L_{shell}

given in Table 6. As shown in Fig. 4, the modelling calculations quantitatively well predicted the experimental results for the leaching of calcium. Figure 7 shows that calcite precipitation was predicted fairly well at lower $NaHCO_3$ concentrations ($\leq 1 \times 10^{-4}$ mol dm^{-3}), though it was underestimated at higher $NaHCO_3$ concentrations. In the CCT-P calculation, the amount of precipitated calcite saturated after 10 ~ 20 weeks owing to clogging; however, in the experiments, calcite still precipitated even if the rate of precipitation decreased with time.

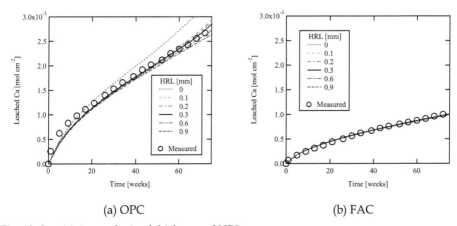

(a) OPC   (b) FAC

Fig. 10. Sensitivity analysis of thickness of HRL

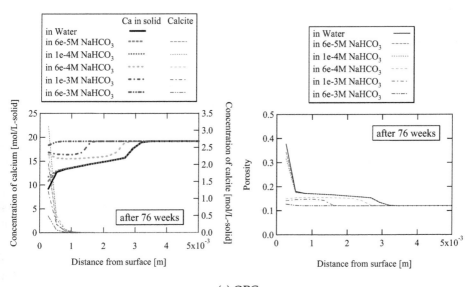

(a) OPC

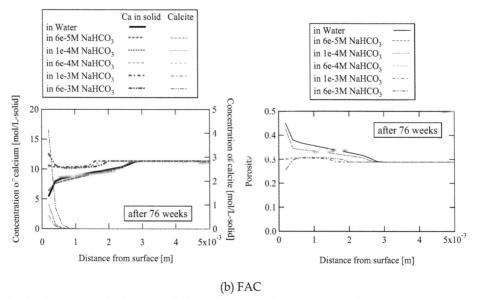

(b) FAC

Fig. 11. Results of calculation modelling alteration of cement monolith

The calculation also predicted qualitatively the calcium concentration in the solid matrix by comparing Fig. 11 with Fig. 6. The porosity in the altered region increased in distilled water, as shown in Fig. 11. The porosity decreased at the surface in NaHCO$_3$ solutions and little alteration in the porosity of the solid was predicted for 1×10^{-3} and 6×10^{-3} mol dm^{-3} NaHCO$_3$ solutions. In the case of FAC, a marked reduction in the porosity in the vicinity of the surface of the solid was predicted owing to calcite precipitation in the region.

5. Discussion

5.1 Mechanism of clogging

Figures 4 and 7 demonstrate that a higher NaHCO$_3$ concentration in a solution induces a larger reduction in the rate of calcium leaching from a cement hydrate monolith due to calcite precipitation. However, in some cases, more calcite did not always precipitate at higher NaHCO$_3$ concentrations; more calcite precipitated at the NaHCO$_3$ concentrations of 6×10^{-4} and 1×10^{-3} mol dm^{-3} than at the NaHCO$_3$ concentration of 6×10^{-3} mol dm^{-3} for OPC, as shown in Fig. 7 (a), and more calcite precipitated at the NaHCO$_3$ concentration of 6×10^{-4} mol dm^{-3} than at the NaHCO$_3$ concentrations of 1×10^{-3} and 6×10^{-3} mol dm^{-3} for FAC, as shown in Fig. 7 (b). These results also suggest that a denser calcite layer was formed at the higher NaHCO$_3$ concentrations and its effect of inhibiting the leaching of calcium was more notable owing to a stronger restriction on mass transport at the interface between the solid and the solution.

The behaviour above can be described using the thickness of the near-surface layer (L_{shell}) as a parameter. In the proposed model, all calcite formed in the solution (in the HRL) precipitates on the near-surface layer so that a denser calcite layer formed as calcite concentrates in a thinner layer (smaller L_{shell}). Kurashige and Hironaga (2007) carried out a series of leaching experiments using OPC and low-heat portland cement in $NaHCO_3$ solution, and found that the secondary calcite precipitation layer grew thinner as the $NaHCO_3$ concentration increased. Based on their finding, it can be assumed that the thickness of the near-surface layer of calcite depends on the $NaHCO_3$ concentration in the solution. This assumption is discussed by the sensitivity analysis of L_{shell}, as shown in Fig. 12. Figure 12 demonstrates that a smaller L_{shell} gives a larger reduction in the rate of calcium leaching. Table 6 gives the best-fit values of the near-surface layer obtained by sensitivity analysis.

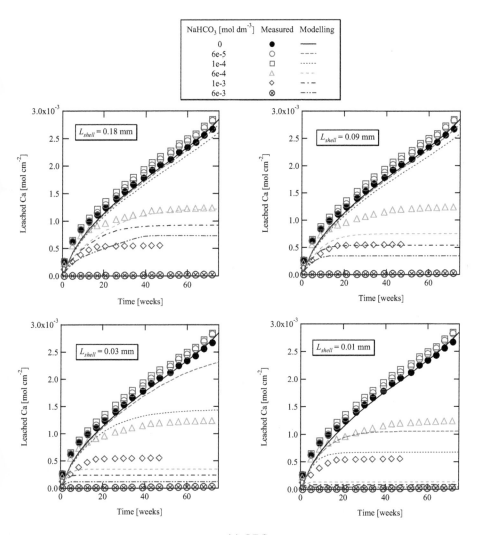

(a) OPC

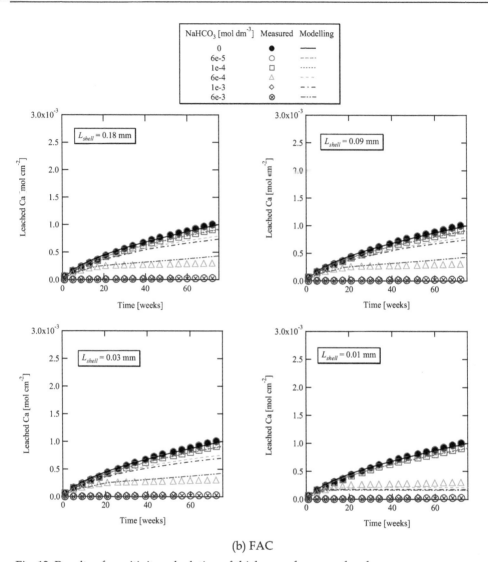

(b) FAC

Fig. 12. Results of sensitivity calculation of thickness of near-surface layer

In the case of FAC, the near-surface layer appears to be insensitive to NaHCO₃ concentration. It could be interpreted that the penetration of carbonate ions into the solid was restricted owing to the low diffusivity in the FAC matrix. It is noted that a lower diffusion coefficient was provided for FAC than for OPC even though the porosity of FAC (29%) was higher than that of OPC (12%). Similar observations were given by Sugiyama et al. (2008) and Chida and Sugiyama (2007, 2008): they observed that the effective diffusion coefficients of uranium (Sugiyama et al., 2008) , organic carbon (Chida and Sugiyama, 2008), caesium, strontium and iodine (Chida and Sugiyama, 2007) in the FAC matrix were more than an order of magnitude lower than those in the OPC matrix, though the porosity of FAC was higher than that of OPC. The reason for the lower diffusion coefficients for FAC is still unclear, but it should be interpreted using the smaller pore size distribution in the FAC matrix (Chida and Sugiyama, 2007). Also, it could be suggested that the pore microstructure (tortuosity of the pore network) affected diffusivity (Promentilla, 2009). Further studies should be carried out to investigate the growth of the secondary precipitation layer to discuss the mechanism of clogging in more detail.

NaHCO$_3$ concentration [mol dm^{-3}]	L_{shell} for OPC [mm]	L_{shell} for FAC [mm]
6 × 10^{-5}	0.18	0.01
1 × 10^{-4}	0.18	0.01
6 × 10^{-4}	0.18	0.01
1 × 10^{-3}	0.09	0.01
6 × 10^{-3}	0.01	0.01

Table 6. Best-fit thickness of near-surface layer (L_{shell})

5.2 Extrapolation of modelling to assess long-term alteration of cementitious repository

It has been demonstrated that the proposed model in this study has the capability of describing the leaching of calcium with the incongruent dissolution of C-S-H and the effect of the precipitation of the secondary less soluble phase (calcite) on the alteration of cement materials. To extrapolate the modelling for the assessment of the long-term alteration of a cementitious repository in the underground repository environment, some issues should be discussed further to improve the reliability of the modelling calculation.

Temporal and spatial discretizations should be carefully optimised in the modelling. A finer mesh size could give a better, more accurate prediction as examined in subsection 4.2; however, the numerical calculation would be instable with a very fine (too small) mesh size, as shown in Fig. 9. It would be impractical to choose very fine spatial and temporal resolutions when considering the computing time for the simulation of the long-term (sometimes up to million years) alteration of an engineering-scale repository (more than tens of meters). A deliberate sensitivity analysis should be carried out to provide a credible assessment. Also, it is suggested that the choice of material could improve the quality of assessment since FAC was less sensitive to the temporal and spatial discretizations, and also the size of HRL according to the modelling results in this study. This could be interpreted on the basis of the low diffusivity in the FAC matrix.

Another important issue is the need to investigate the mechanism of clogging in more detail. In the underground environment, cementitious materials are placed in contact with groundwater through other engineering barrier materials or surrounding rocks. The rate of alteration could be moderated compared to the experiment in which the cement solid is placed in contact with the solution directly. Further experimentation is required on the alteration on the solid in an actual repository environment and enables us to establish optimised calculation conditions including spatial and temporal resolutions, the appropriate HRL, and the thickness of the near-surface layer (L_{shell}). In situ pilot experiments in the repository environment will be effective in tackling this issue.

6. Conclusions

A reactive transport computational code, in which a geochemical model including the thermodynamic incongruent dissolution model of C-S-H is coupled with the advection-diffusion/dispersion equation, was developed on the basis of a series of experiments on the alteration of hydrated cement monoliths in deionised water and in sodium bicarbonate solution. The code can describe the evolution of the hydraulic properties of the solid cement matrix due to the leaching and precipitation of components and the clogging effect by insoluble secondary phase precipitation that inhibits the alteration of cement materials.

The precipitation of secondary calcite on the surface of cement solid induces a reduction in the rate of components leaching. The clogging behaviour depends on the feature of the near-surface layer of precipitates, which is sensitive to the concentration of reactive ions in the solution and the mass transport property of the solid matrix. Further experimental studies are needed to analyse in detail the alteration at the interface between cementitious material and other barrier materials (e.g., bentonite and host rock) under actual underground repository conditions. FAC, in which a low diffusivity is given, is suggested to reduce the uncertainty in the long-term performance assessment.

7. References

Atkinson, A. (1985). *The Time Dependence of pH within a Repository for Radioactive Waste Disposal*, AERE R 11777, UKAEA, 1985.

Atkinson, A.; Goult, D.J. & Hearne, J.A. (1985). An Assessment of the Long-term Durability of Concrete in Radioactive Waste Repositories, *Scientific Basis for Nuclear Waste Management IX (Material Research Society Symposium Proceedings volume 50)*, pp. 239-246, Stockholm, Sweden, September 1985.

Bond, K.A.; Heath, T.G. & Tweed, C.J. (1997). *HATCHES: A Referenced Thermodynamic Database for Chemical Equilibrium Studies*, Nirex Report NSS/R379, 1997.

Brodersen, K. (2003). *CRACK2 – Modelling Calcium Carbonate Deposition from Bicarbonate Solution in Cracks in Concrete*, Risø-R-1143(EN), ISBN 87-550-2612-5, 2003.

Burnol, A.; Blanc, P.; Tournassat, C.; Lassin, A.; Xu, T. & Gaucher, E.C. (2005). Intercomparison of Reactive Transport Models Applied to Degradation of a Concrete/Clay Interface, *Migration '05*, Avignon, France, September 2005.

Chida, T. & Sugiyama, D. (2007). Observation of Diffusion Behavior of Trace Elements in Hardened Cement Pastes by LA-ICP-MS, *Scientific Basis for Nuclear Waste Management XXXI (Material Research Society Symposium Proceedings volume 1107)*, pp. 585-592, ISBN 978-1-60511-079-0, Sheffield, UK, September 2007.

Chida, T. & Sugiyama, D. (2008). Diffusion Behavior of Organic Carbon and Iodine in Low-heat Portland Cement Containing Fly Ash, *Scientific Basis for Nuclear Waste Management XXXII (Material Research Society Symposium Proceedings volume 1124)*, pp. 379-384, ISBN 978-1-60511-096-7, Boston, USA, December 2008.

Glasser, F.P.; Macphee, D.E. & Lachowski, E.E. (1987). Modelling Approach to the Prediction of Equilibrium Phase Distribution in Slag-cement Blends and Their Solubility Properties, *Scientific Basis for Nuclear Waste Management XI (Material Research Society Symposium Proceedings volume 112)*, pp. 3-12, ISBN 0-931837-82-0, Boston, USA, December 1987.

Glasser, F.P.; Adenot, F.; Bloem Bredy, P.J.C.; Fachinger, J.; Sneyers, A.; Marx, G.; Brodersen, K.; Cowper, M.; Tyrer, M. et al. (2001). *Barrier Performance of Cements and Concretes in Nuclear Waste Management*, Final Report for CEC Contract FI4W-CT96-0030, EUR 19780 EN, 2001.

Haga, K.; Sutou, S.; Hironaga, M.; Tanaka, S. & Nagasaki, S. (2005). Effects of Porosity on Leaching of Ca from Hardened Ordinary Portland Cement Paste, *Cement and Concrete Research*, Volume 35, Issue 9, (September 2005), pp. 1764-1775, ISSN 0008-8846.

Harris, A.W.; Atkinson, A.; Balek, V.; Brodersen, K.; Cole, G.B.; Haworth, A.; Malek, Z.; Nickerson, A.K.; Nilsson, K. & Smith, A.C. (1998). *The Performance of Cementitious Barriers in Repositories*, Final Report for CEC Contract FI2W-0040, EUR 16643 EN, 1998.

Hindmarsh, A.C. & Gresho, P.M. (1984). The Stability of Explicit Euler Time-Integration for Certain Finite Difference Approximations of the Multi-Dimensional Advection-Diffusion Equation, *International Journal for Numerical Methods in Fluids*, Volume 4, Issue 9, (September 1984), pp. 853-897, ISSN 1097-0363.

Kurashige, I. & Hironaga, M. (2007). *Mechanism of Leaching Inhibition of Cementitious Materials due to Hydrogencarbonate Ion in Groundwater*, CRIEPI Report N06028, April 2007 [in Japanese with English Abstract], ISBN 4-86216-502-8.

Lagneau, V. & van der Lee, J. (2005). Simulation of Clogging Effects at the Interface between MX80-clay and Concrete in a Deep Radioactive Waste Repository, *Migration '05*, Avignon, France, September 2005.

Marty, N.C.M.; Tournassat, C.; Burnol, A.; Giffaut, E. & Gaucher, E.C. (2009). Influence of Reaction Kinetics and Mesh Refinement on the Numerical Modelling of Concrete/Clay Interactions, *Journal of Hydrology*, Volume 364, Issues 1-2, (15 January 2009), pp. 58-72, ISSN 0022-1694.

Parkhurst D.L. et al. (1980). *PHREEQE - A Computer Program for Geochemical Calculations*, U.S. Geological Survey.

Promentilla, M.A.B.; Sugiyama, T.; Hitomi, T. & Takeda, N. (2009). Quantification of Tortuosity in Hardened Cement Pastes Using Synchrotron-based X-ray Computed Microtomography, *Cement and Concrete Research*, Volume 39, Issue 6, (June 2009), pp. 548-557, ISSN 0008-8846.

Sousa, E. (2003). The Controversial Stability Analysis, *Applied Mathematics and Computation*, Volume 145, Issues 2-3, (25 December 2003), pp. 777–794, ISSN 0096-3003.

Sugiyama, D. (2008). Chemical Alteration of Calcium Silicate Hydrate (C–S–H) in Sodium Chloride Solution, *Cement and Concrete Research*, Volume 38, Issue 11, (November 2008), pp. 1270-1275, ISSN 0008-8846.

Sugiyama, D. & Fujita, T. (2006). A Thermodynamic Model of Dissolution and Precipitation of Calcium Silicate Hydrates, *Cement and Concrete Research*, Volume 36, Issue 2, (February 2006), pp. 227-237, ISSN 0008-8846.

Sugiyama, D.; Fujita, T.; Chida, T. & Tsukamto, M. (2007). Alteration of Fractured Cementitious Materials, *Cement and Concrete Research*, Volume 37, Issue 8, (August 2007), pp. 1257-1264, ISSN 0008-8846.

Sugiyama, D.; Chida, T. & Cowper, M. (2008). Laser Ablation Microprobe Inductively Coupled Plasma Mass Spectrometry Study on Diffusion of Uranium into Cement

Materials, *Radiochimica Acta*, Volume 96, Issue 9-11, Migration 2007, pp. 747-752, ISSN 0033-8230.

Taylor, H.F.W. (1997). *Cement Chemistry 2nd ed.*, Thomas Telford Services Ltd., p. 224, ISBN 0-7277-2592-0, London.

TRU Coordination Office (Japan Nuclear Cycle Development Institute and The Federation of Electric Power Companies) (2000). *Progress Report on Disposal Concept for TRU Waste in Japan*, JNC TY1400 2000-002, TRU TR-2000-02, March 2000.

Yasuda, K.; Yokozeki, K.; Kawata, Y. & Yoshizawa, Y. (2002). Physical and Transportation Properties of Concrete due to Calcium Leaching, *Cement Science and Concrete Technology*, Japan Cement Association, Volume 56, pp. 492-498, ISSN 0916-3182. [in Japanese with English Abstract].

7

Clarification of Adsorption Reversibility on Granite that Depends on Cesium Concentration

Keita Okuyama and Kenji Noshita
Hitachi Research Laboratory, Hitachi, Ltd.
Japan

1. Introduction

Disposal of high-level radioactive wastes (HLW) is planned to be done in a repository located deep underground to isolate radionuclides from the biosphere. In case of a leakage accident of HLW, there will be no hazardous impact to humans because migration of the leaked radionuclides will be retarded by matrix diffusion and adsorption on the rock surface. Therefore, the geochemical retardation behavior of radionuclides in aquifers must be clarified, from the viewpoint of the performance assessment of HLW deep underground disposal.

Radionuclide-adsorbed sites in rock are classified into two general types: reversible adsorption sites, where desorption of once-adsorbed nuclides occurs (e.g., an ion-exchange reaction); and irreversible adsorption sites, where only adsorption occurs (e.g., a mineralization reaction). In the early stage of radionuclide leakage, migration of the radionuclides will be retarded by both reversible and irreversible adsorption sites. However, when the irreversible sites are filled, the radionuclides will no longer be retarded by them. Therefore, it is necessary to investigate the sorption reversibility to clarify the behavior of radionuclides (Fukui, 2004).

Cesium-137 (^{137}Cs), which is one of the principal radioactive sources of HLW for 1000 years after geological disposal, is partially fixed in the interlayer of micas and it might be trapped irreversibly in these sites (Francis & Brinkley, 1976). Another report suggested that Cs adsorption on granite has a nonlinear relationship with its concentration in solution (Ohe, 1984). These findings imply it is possible to characterize Cs adsorption on granite. One of the important characterization factors is the adsorption amount on reversible and irreversible sites, which may change with Cs concentration; however, it has not been reported.

The purpose of this study is to clarify Cs adsorption reversibility on granite that depends on Cs concentration by adsorption and sequential extraction experiments, using a variety of chemical reagents for various inlet Cs concentrations (1.0×10^{-3} - 1.0×10^{0} mol/m^3). For the experiments, a narrow flow channel was formed on a granite specimen (Okuyama, et al., 2008). Breakthrough curves (BTCs) were obtained by injecting a Cs solution labeled with

[134]Cs into the channel. We estimated the amounts diffused into the rock matrix and absorbed on the rock surface. In order to verify the estimated sorption amount, we obtained the retardation coefficient by analyzing BTCs, and compared the value with that calculated using the distribution coefficient (Kd) obtained by a batch sorption experiment. After injecting [134]Cs solution, extraction reagents (HCl, $CaCl_2$, and KCl solutions) were injected into the flow channel in sequence. We obtained the desorption curves and investigated the chemical speciation of Cs in granite. In particular, we focused on the dependence of adsorption amount on reversible and irreversible sites on Cs concentration.

2. Experiment

2.1 Materials

Biotite granite from the Makabe area of Japan was used in this work. The chemical composition was measured by fluorescent X-ray analysis and the results are listed in Table 1. The mineral composition was also graphically determined using a polarization microscope and these results are listed in Table 2. The porosity of a granite specimen was measured by the water saturation method (Skagius & Neretnieks, 1986), and 0.73 % was obtained. The cation-exchange capacity was determined by the Peech method (Peech, 1965), and 5.9 meq/100g was obtained.

2.2 Adsorption and sequential extraction experiments

In order to clarify sorption reversibility of Cs on granite, we carried out adsorption and sequential extraction experiments. All specimens were cut as a parallelepiped with dimensions of 20 mm × 4 mm × 3 mm and their surfaces were polished with abrasive paper (#1200). The side edge surfaces were thickly coated with an epoxy resin to avoid evaporation of the test solution during experiments (Fig.1). A container was filled with distilled water (pH adjusted to 6 using HCl and NaOH) and set in a glass bell jar. Granite specimens were immersed in the water. Then, the bell jar was evacuated to remove air from the micropores of the specimens. The specimens were kept for 24 hours immersed in the distilled water in the bell jar under reduced pressure.

Experimental apparatus is shown in Fig.2. It consisted of an injection unit, a reaction unit, and a storage unit. For the reaction unit, a narrow flow channel (20 mm length, 4 mm width, and 160 μm depth) was formed from a fluoroplastic spacer (160 μm thick) with a slit (20 mm long, 4 mm wide). Two fluoroplastic base plates sandwiched the spacer and granite specimen, then everything was held together by applying pressure to downward from the top base plate and upward from the bottom base plate. A photograph of the spacer is shown in Fig.3. The roughness of the granite specimen and the bottom base plate surfaces was overcome by elastic deformation of the fluoroplastic spacer, so no leakage of solution occurred. The bottom base plate had inlet and outlet ports (0.5 mm inner diameter), for injecting or draining out radionuclide or extraction solutions, and fluoroplastic tubing was attached to each port. A blank experimental run in which the granite specimen was replaced with a fluoroplastic plate was carried out to verify the surfaces of the fluoroplastic spacer and base plates were non-reactive, then the actual experiments were done using the following procedures.

Oxide	Content (wt%)
SiO$_2$	69.2
Al$_2$O$_3$	15.9
K$_2$O	4.30
CaO	2.97
Na$_2$O	2.90
Fe$_2$O$_3$	2.54
P$_2$O$_5$	0.60
MgO	0.48
TiO$_2$	0.29
MnO	0.05
Total	99.24

Table 1. Granite chemical composition

Oxide	Content (wt%)
Quartz	37.9
Plagioclase feldspar	33.0
Potassium feldspar	18.8
Biotite	9.0
Chlorite	1.0
Prehnite	0.2
Carbonate mineral	0.1
Total	100.0

Table 2. Granite mineral composition

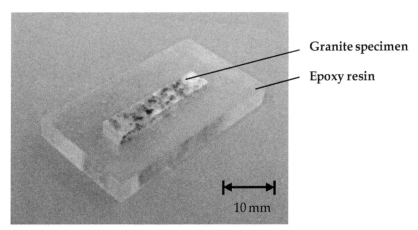

Fig. 1. Photograph of a granite specimen showing side edge surfaces coated with an epoxy resin

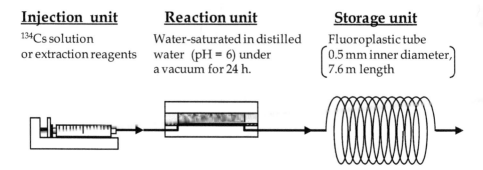

Injection unit

^{134}Cs solution
or extraction reagents

Reaction unit

Water-saturated in distilled
water (pH = 6) under
a vacuum for 24 h.

Storage unit

Fluoroplastic tube
$\left[\begin{array}{l}0.5\,\text{mm inner diameter,}\\ 7.6\,\text{m length}\end{array}\right]$

Detail of reaction unit

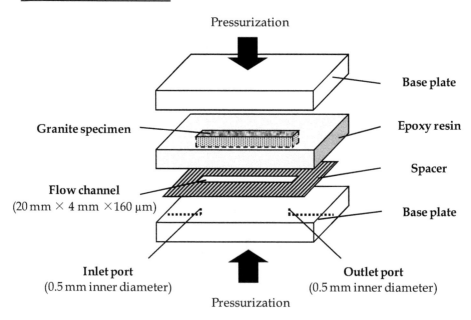

Pressurization

Base plate

Granite specimen

Epoxy resin

Spacer

Flow channel
(20 mm × 4 mm ×160 μm)

Base plate

Inlet port
(0.5 mm inner diameter)

Outlet port
(0.5 mm inner diameter)

Pressurization

Fig. 2. Experimental apparatus for adsorption and sequential extraction experiments

In an experimental run, Cs solution labeled with ^{134}Cs (specific activity adjusted to 9.8×10^8 Bq/m^3) was injected into the flow channel at constant flow velocity (2.6×10^{-5} m/s). The flow velocity was maintained for 24 hours by using an injection pump in all experiments. Average residence time in the flow channel was 12.8 minutes. To avoid evaporation of effluent solution during fractionation, effluent solution was stored in a storage unit (0.5 mm inner diameter, 7.6 m length). After an adsorption experimental run, the stored solution was flushed out with a high flow velocity (1.3×10^{-3} m/s), and collected in a small vial. Collection quantity was obtained precisely by weighing the vials before and after runs. The concentrations of ^{134}Cs in the effluent of the flow channel were determined with a germanium (Ge) semiconductor detector. The error of concentration measurement was less than 5%. The experimental apparatus, excluding the radioactivity detectors, was assembled in a glove box filled with air to keep dust particles off and all experimental runs were done at 25 ± 2 °C. After these procedures were carried out, extraction reagents were injected (Table 3) into the flow channel in sequence under the same operating conditions for the sequential extraction experiments. A series of experimental runs was carried out by changing the Cs concentration (1.0×10^{-3} - 1.0×10^{0} mol/m^3).

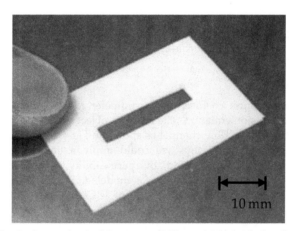

Fig. 3. Photograph of a fluoroplastic thin spacer (160 μm thick) with the slit which forms the flow channel on the granite plate

Extractant	Composition (mol/m³)	pH
HCl	1.0×10^{-2}	5
CaCl$_2$	5.0×10^{2}	5
KCl	5.0×10^{2}	5

Table 3. Extraction reagents

2.3 Batch sorption experiment

We estimated Cs adsorption amount on granite by BTCs. In order to verify this estimation, we evaluated the retardation coefficient by analyzing BTCs, and compared it with the value calculated using the distribution coefficient (Kd) obtained by a batch sorption experiment. The experimental procedures for the batch sorption experiment were as follows. A granite specimen was crushed and sieved to obtain particle sizes in the range of 0.125 - 0.25 mm.

The sieved particles were rinsed with distilled water to remove the fine powder fragments. Specific surface area was measured by the BET method, and the mean value was 1.0×10^{-2} m^2/kg. Then three grams of the sieved particles was saturated with distilled water (3.0×10^{-5} m^3, adjusted to pH = 6) spiked with a radionuclide solution of ^{134}Cs (specific activity 4.9×10^7 Bq/m^3); the saturation was done under a vacuum for 24 hours to fill the granite particle pores. The initial Cs$^+$ solution concentrations ranged from 1.0×10^{-4} mol/m^3 to 1.0×10^2 mol/m^3. The particle-containing solutions were continuously stirred for seven days. Then the solid particles were removed by filtration through a membrane filter (pore size 0.45 µm) and the ^{134}Cs concentration was measured with the Ge semiconductor detector. The Kd value was calculated by the conventional procedure as follows (Holland and Lee, 1992):

$$Kd = \frac{C_{in} - C_{eq}}{C_{eq}} \cdot \frac{v}{w} \tag{1}$$

where C_{in} is initial radionuclide concentration (mol/m^3), C_{eq} is equilibrium radionuclide concentration (mol/m^3), v is solution volume (m^3) and w is rock (granite) weight (kg).

3. Analysis

We obtained the retardation coefficient by parameter identification method for BTCs using an advection-diffusion equation. The analysis object was a vertical section through the granite specimen and the flow channel as shown in Fig.4. In the calculation, the flow channel was not deep (160µm); thus we could apply a one-dimensional advection-dispersion equation. On the other hand, the permeability of granite was much small thus a two-dimensional diffusion equation was modeled. The test time was several days at the longest; therefore the radioactive decay of ^{134}Cs (half life 2.07 years) could be neglected.

The governing equations were thus formulated as follows:

For the flow channel

$$R_f \frac{\partial C_f}{\partial t} = -V \frac{\partial C_f}{\partial x} + D_L \frac{\partial^2 C_f}{\partial x^2} + \frac{D_e}{b} \frac{\partial C_m}{\partial z}\bigg|_{z=0} \tag{2}$$

$$(t > 0, \; 0 < x < L)$$

For the granite matrix region

$$R_m \frac{\partial C_m}{\partial t} = \frac{D_e}{\varepsilon} \frac{\partial^2 C_m}{\partial x^2} + \frac{D_e}{\varepsilon} \frac{\partial^2 C_m}{\partial z^2} \tag{3}$$

$$(t > 0, \; 0 < x < L, \; 0 < z < d)$$

where, b is depth of the flow channel (m), $C_f(x,t)$ is radionuclide concentration in the flow channel (mol/m^3), $C_m(x,z,t)$ is radionuclide concentration in the porous granite (mol/m^3), d

is thickness of granite specimen (m), D_e is effective diffusion coefficient (m²/s), D_L is longitudinal dispersion coefficient for the flow channel (m²/s), ε is porosity of granite (-), L is length of the flow channel (m), R_f is retardation factor of granite surface (-), R_m is retardation factor of granite matrix (-), t is elapsed times (s), V is flow velocity in the flow channel (m/s), x is distance along the flow channel (m) and z is vertical distance from the flow channel (m).

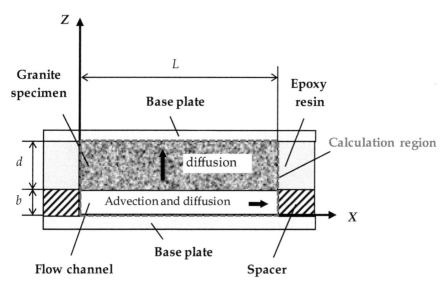

Fig. 4. Analysis object; vertical section showing the granite plate and flow channel

The boundary conditions around the flow channel were as shown in Eqs. (4) and (5):

$$VC_f - D_L \left.\frac{\partial C_f}{\partial x}\right|_{x=0} = \frac{F}{A}, \quad (t>0) \tag{4}$$

$$C_m(x,0,t) = C_f(x,t), \quad (0<x<L, \quad t>0) \tag{5}$$

where, F is the inlet flux of radionuclide (mol/s) and A is cross sectional area of the flow channel (m²).

Other boundary conditions around the granite specimen were as shown in Eqs. (6) – (8):

$$\frac{\partial C_m(x,d,t)}{\partial z} = 0, \quad (0<x<L, \ t>0) \tag{6}$$

$$\frac{\partial C_m(0,z,t)}{\partial x} = 0, \quad (0<z<d, \ t>0) \tag{7}$$

Symbol	Parameter	Value
A	Cross sectional area of the flow channel	6.4×10^{-7} m²
b	Depth of the flow channel	1.6×10^{-4} m
d	Thickness of granite specimen	3.0×10^{-3} m
D_e	Effective diffusion coefficient	1.5×10^{-11} m
D_L	Longitudinal dispersion coefficient for the flow channel	0 m²/s
D_{Lst}	Longitudinal dispersion coefficient for the storage unit	3.0×10^{-5} m²/s
ε	Porosity of granite	0.73 %
L	Length of the flow channel	2.0×10^{-2} m
L_{st}	Length of the storage unit	7.6 m
t	Elapsed times	86400 s.
V	Flow velocity in the flow channel	2.6×10^{-5} m/s
V_{st}	Flow velocity in the storage unit	1.3×10^{-3} m/s

Table 4. BTCs calculation condition and parameters

$$\frac{\partial C_m(L, z, t)}{\partial x} = 0, \quad (0 < z < d, \ t > 0) \tag{8}$$

The calculation parameters are given in Table 4. The effective diffusion coefficient (D_e) was related to the diffusion coefficient in free water (D_0) and the formation factor (FF) (Skagius, et al., 1982) as given by Eq.(9):

$$D_e = \varepsilon \cdot \frac{\delta}{\tau^2} \cdot D_0 = FF \cdot D_0 \tag{9}$$

where, δ is constrictivity (-) and τ^2 is tortuosity (-). The value of FF for ³H was used instead of that for ¹³⁴Cs because non-sorbing characteristics of ³H lead to direct determination of FF as shown in Eq. (10):

$$FF = \left(\frac{D_e}{D_0}\right)_{3H} \tag{10}$$

The values of D_e and D_0 for ³H are 1.5×10^{-11} m²/s (Okuyama, et al., 2008) and 2.00×10^{-9} m²/s (Chemical Society of Japan, 2004), respectively. By substituting the values of D_0 for Cs as 2.04×10^{-9} m²/s (Chemical Society of Japan, 2004) into Eq. (9), we obtained $D_e = 1.5 \times 10^{-11}$ m²/s for Cs; we used this value for BTC analysis.

When the Cs-containing solution flowed through the flow channel and storage unit, it was dispersed in the longitudinal direction. The length of the flow channel (20 mm) was much shorter than that of the storage unit (7.6 m length), thus we neglected dispersion in the flow channel. In order to obtain longitudinal dispersion coefficient for the storage unit D_{Lst}, we calculated the value of the dispersion length a by analyzing the one-dimensional advection-dispersion equation as Eqs. (11) and (12):

$$\frac{\partial C_{st}}{\partial t} = -V_{st}\frac{\partial C_{st}}{\partial x} + D_{Lst}\frac{\partial^2 C_{st}}{\partial x^2} \tag{11}$$

$$\left(t > 0, \quad 0 < x < L_{st}\right)$$

$$D_{Lst} = D_0 + a \cdot V_{st} \tag{12}$$

where, a is dispersion length (m). $C_{st}(x, t)$, L_{st} and V_{st} are radionuclide concentration (mol/m^3), length (m) and flow velocity (m/s), in the storage unit, respectively.

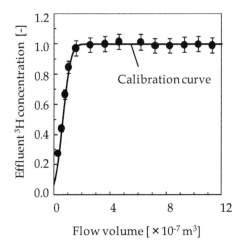

Fig. 5. Experimental BTC values of ^{3}H with the calibration curve using the results of the numerical analysis

The BTC of ^{3}H, which is a non-sorbing species, is shown in Fig. 5. In a blank experimental run, we confirmed that ^{3}H was not absorbed on the fluoroplastic tubing, thus the outlet concentration agreed with the inlet concentration (C_{st} / C_0 = 1). The transient region to the value of C_{st} / C_0 = 1 represented the effect of dispersion in the storage unit. The value of a was determined by a parameter identification method and a = 0.023 m was obtained. By substituting the values of a into Eq. (12), we obtained D_{Lst} = 3.0 ×10^{-5} m^2/s for storage unit; thus this value was used for BTC analysis. Equations (2) and (3) were solved, subject to Eqs.(4)-(8), using a finite volume scheme.

4. Results and discussion

4.1 Adsorption of ^{134}Cs

We evaluated Cs adsorption amount on granite by BTCs. BTCs was obtained by injecting the Cs solution labeled with ^{134}Cs into the flow channel at constant flow velocity. Figure 6 plots BTCs of ^{134}Cs for various inlet Cs concentrations, showing the change in normalized effluent concentration C/C_0, which is the effluent concentration C divided by the inlet concentration C_0, as a function of elapsed time. A quasi-plateau region was observed for each curve and the breakthrough values in this region did not reach the equilibrium $C/C_0 = 1$. This is because some of the Cs may be still driven away by diffusion into the granite matrix or they were absorbed on the rock surface. From the mass balance standpoint, we assumed that the area enclosed by the lines of $C/C_0 = 1$ and the BTC represented the Cs diffusion into the granite matrix or the Cs absorbed on the granite surface.

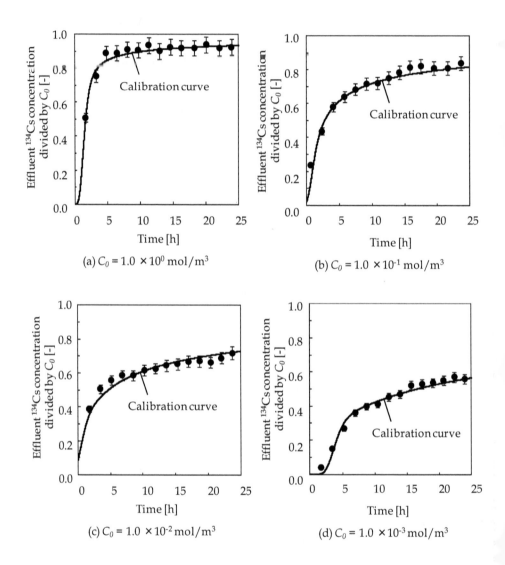

(a) $C_0 = 1.0 \times 10^0$ mol/m^3

(b) $C_0 = 1.0 \times 10^{-1}$ mol/m^3

(c) $C_0 = 1.0 \times 10^{-2}$ mol/m^3

(d) $C_0 = 1.0 \times 10^{-3}$ mol/m^3

Fig. 6. Experimental BTCs of ^{134}Cs for various inlet Cs concentrations with a curve using the result of the numerical analysis

In order to verify this estimation of sorption amount, we determined the retardation coefficient by parameter identification method for BTCs using Eqs. (2), (3), and (11), and compared it with the retardation coefficient (R_m) calculated using the value of Kd obtained by a batch sorption experiment. By inserting ε, Kd, and density of granite ρ (2.98 ×10³ kg/m³) into Eq.(13), R_m were calculated.

$$R_m = 1 + \frac{\rho \cdot Kd}{\varepsilon} \tag{13}$$

The R_m values obtained by the two methods were almost identical, therefore the area enclosed by the lines of $C/C_0 = 1$ and BTC represented the Cs diffusion into the granite matrix or the Cs absorbed on the rock surface (Okuyama, et al., 2008).

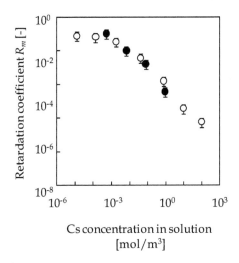

● Analyzing BTC values

○ Batch sorption experiment results

Cs concentration in solution [mol/m³]

Fig. 7. Retardation coefficients obtained by analyzing BTC values and batch sorption experiment results

4.2 Desorption of ^{134}Cs

After injecting ^{134}Cs solution, extraction reagents (HCl, CaCl₂, and KCl solutions as shown in Table 3) were injected into the flow channel in sequence. We obtained the desorption curves and investigated the chemical speciation of Cs in granite, in particular the ratio of the density of reversible and irreversible adsorption sites depends on Cs concentration in solution. Desorption curves are shown in Fig. 8 as the change in normalized desorbed Cs divided by the total adsorption amount. For each curve, desorption amount of Cs was significantly decreased to 1 % or lower when elapsed time was 24, 48, and 72 hours, so Cs was adequately desorbed by each reagent. Desorption amount of Cs changed drastically between the reagents, thus it was conceivable that desorbed Cs for each reagent indicated the presence of multiple adsorption sites.

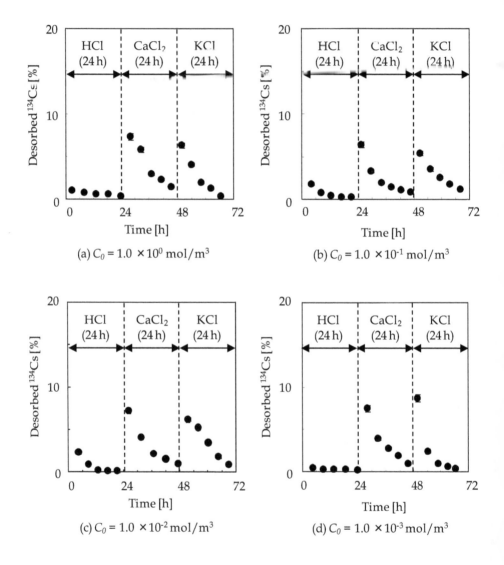

Fig. 8. Experimental desorption curves of [134]Cs for various inlet Cs concentrations

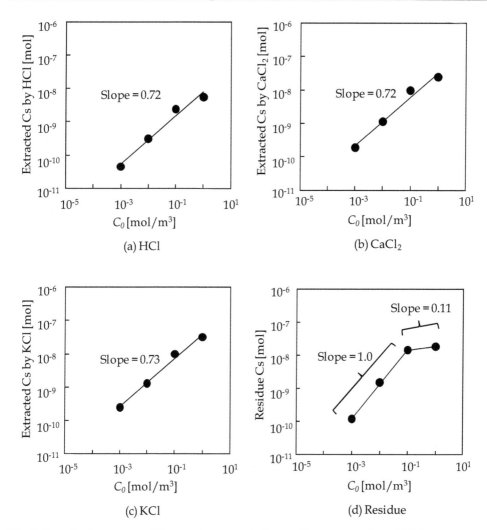

Fig. 9. Desorbed amounts of Cs at various quasi-plateau Cs concentration of BTCs

Desorption amount of Cs at various quasi-plateau Cs concentration of BTCs is summarized in Fig. 9. The fractions extracted with HCl, KCl and $CaCl_2$ gave straight lines and the slopes were about 0.7, which could be described by a Freundlich isotherm. On the other hand, the residue fraction was directly proportional to the Cs concentration at low Cs concentration (lower than 1×10^{-1} mol/m^3), and showed signs of leveling off at 1×10^0 mol/m^3; thus Cs adsorption on granite surface may become saturated at Cs concentration of 1×10^0 mol/m^3.

The chemical speciation of Cs desorbed by each regent was described as follows. Biotite has two types of adsorption sites; variable-charge sites, and permanent-charge sites. The charge of variable-charge sites changes with changing pH, thus the fraction of Cs extracted with HCl indicates adsorption amount on variable-charge sites. On the other hand, it has been reported that Cs is strongly absorbed on biotite grains; in particular it is distributed onto the interlayers of these grains[2], which is permanent-charge sites. The hydrated ionic radius of K^+ (0.13 nm) (Ohtaki, 1990) are almost identical to that of Cs^+ (0.12 nm) (Ohtaki, 1990), thus Cs^+ which has been taken into the interlayers induces interlayer spacing that is large enough to permit diffusion of K^+; thus K^+ would displace adsorbed Cs on the biotite interlayer permanent-charge sites. The hydrated ionic radius of Ca^{2+} (0.31 nm) (Ohtaki, 1990) is much larger than that of Cs^+ and Ca^{2+} diffusion into the interlayers is restricted (Cornell, 1993); thus Ca^{2+} would displace adsorbed Cs on the edge of the biotite interlayer or at other mineral sites such as feldspar (Brown, et al., 1984). From the above, although the chemical speciation of Cs desorbed for the three regents was different, the adsorption mechanism of Cs for each was an ion-exchange reaction. The adsorption mechanism of the residue fraction might be fixed in biotite[2]. In this study, we defined that as irreversible adsorption.

The fraction of Cs desorbed by the three reagents is shown in Fig. 10. The sum of fractions extracted with HCl, $CaCl_2$ and KCl, which indicated reversible adsorption, was 60 - 80 %. Irreversible adsorption was relatively large at high Cs concentration. However, Cs adsorption was saturated at 1×10^0 mol/m^3, and the fraction of irreversible adsorption decreased.

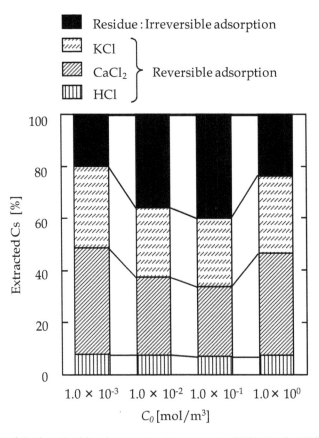

Fig. 10. Fraction of Cs desorbed by three extraction reagents (HCl, $CaCl_2$, KCl)

When the irreversible sites are saturated, the Cs will no longer be retarded by them. Thus, it is concluded that the ratio of the density of reversible and irreversible adsorption sites must be thoroughly considered to demonstrate the safety of HLW deep underground disposal option.

5. Conclusion

We carried out adsorption and sequential extraction experiments using three chemical reagents (HCl, CaCl$_2$, and KCl solutions) for various inlet Cs concentrations to investigate the chemical speciation of Cs in granite. In particular, we focused on the dependence of adsorption amount on reversible and irreversible sites on Cs concentration. The sum of fractions extracted with HCl, CaCl$_2$ and KCl, which indicated reversible adsorption, was 60 - 80 %. Irreversible adsorption was relatively large at high Cs concentration. However, Cs adsorption became saturated, and the fraction of irreversible adsorption decreased. Cs adsorption on granite surface was saturated at Cs concentration of 1×10^0 mol/m^3. In the early stage of radionuclide leakage, Cs will be retarded by the presence of reversible and irreversible adsorption sites. However, when the irreversible sites were saturated, Cs will not be retarded by them. Thus, the sorption reversibility must be thoroughly considered to demonstrate the behavior of Cs in granite, in particular for high Cs concentration.

6. References

Brown, D. L., Haines, R. I., Owen, D. G., Stanchell, F. W. & Watson, D. G. (1984). Surface studies of the interaction of cesium with feldspars. *American chemical society*, 246, pp. 217-227

Cornell, R. M. (1993). Adsorption of cesium on minerals: A review. *J. Radioanalytical and Nuclear Chemistry*, 171, 2, pp. 483-500

Francis, C. W. & Brinkley F. S. (1976). Preferential adsorption of ^{137}Cs to micaceous minerals in contaminated freshwater sediment. *Nature*, 260, pp. 511-513

Fukui, M. (2004). Affinity of trace elements to sandy soil and factors affecting on the migration through media. *KURRI KR*, 99, pp. 21-26

Holland, T. R. & Lee, D. J. (1992). Radionuclide getters in cement. *Cement and Concrete Research*, 22, pp. 247-258

Chemical Society of Japan. (Ed.). (2004). *Kagaku Binran, Basic 5th ed.,* Maruzen Company, Tokyo, pp. 64-65

Ohe, T. (1984). Ion exchange adsorption of radioactive cesium, cobalt, manganese, and strontium to granitoid rocks in the presence of competing cations. *J. Nucl. Sci. Technol.*, 67, 1, pp. 92-101

Ohtaki, H. (1990). Hydration of Ions. *Kyoritsu Shuppan*, Tokyo, pp. 55

Okuyama, K., Sasahira, A., Noshita, K., & Ohe, T. (2008). A method for determining both diffusion and sorption coefficients of rock medium within a few days by adopting a micro-reactor technique. *Applied Geochemistry*, 23, pp. 2130-2136

Peech, M. (1965). Exchange acidity: Barium chloride-triethanolamine method. *Methods of soil analysis, Part 2*, American Society of Agronomy, Madison, Wisconsin, pp. 910–911

Skagius, K. & Neretnieks, I. (1982). Diffusion in crystalline rock. *Nucl. Waste Manage.*, 5, pp. 509–518

Skagius, K. & Neretnieks, I. (1986). Porosities and diffusivities of some non-sorbing species in crystalline rocks. *Water Resour. Res.*, 22, pp. 389-398

Part 3

Municipal Waste Disposal at Different Environments

Assessment of Population Perception Impact on Value-Added Solid Waste Disposal in Developing Countries, a Case Study of Port Harcourt City, Nigeria

Iheoma Mary Adekunle[1,*], Oke Oguns[1], Philip D. Shekwolo[1],
Augustine O. O. Igbuku[1] and Olayinka O. Ogunkoya[2]
*[1]Remediation Department,
Shell Petroleum Development Company (Nigeria) Limited, Port Harcourt,
[2]University Liaison,
Shell Petroleum Development Company (Nigeria) Limited, Port Harcourt,
Nigeria*

1. Introduction

1.1 Background

Waste materials are solid, liquid, gaseous or radioactive substances that have lost value to the user. Waste materials are often produced by human activities and waste management is the collection, transportation, processing, recycling or disposal, sequestration and monitoring of waste materials; a step undertaken to reduce their effect on health and the environment or aesthetics. Waste management is also carried out to recover resources from waste materials. Waste management is executed via different methods and fields of expertise for each form of waste material. Just as the manufacture, distribution and uses of products result in the emission of greenhouse gases (GHGs) that promote global warming and climate change, improper disposal of the waste materials generated from manufactured products could also promote climate change through the emission of GHGs. Increase in waste material generation is attributed to accelerated population growth, industrialization and urbanization.

Across board, the majority of municipal solid wastes consists of biodegradable organic substances, plastics, glass, metals, textiles and rubber materials but the composition and volume of the wastes vary from one region to the other and also from one country to another. There are also differences between the waste composition of high and low-income countries. In developing countries, waste generation rates are put at 0.66 kg/capita/day in urban areas and 0.44 kg/capita/day in rural areas as opposed to 0.7-1.8 kg/capita/day in developed countries (Cointreau, 1982). Wastes materials in developing countries are characterized by increased putrescible /organic matter contents in comparison to the

---

* Corresponding Author

developed countries; waste stream is over 50% organic materials; residential and market wastes are characterized by 78 to 90% compostable materials (Hoornweg et al., 1999; Cointreau, 1982; Ogwueleka, 2009; Otti, 2011). Furthermore, waste densities and moisture are much higher in developing countries. For instance, the density of solid waste in Nigeria is reported to be 250 kg/m³ to 370 kg/m³ higher than solid waste densities found in developed countries (Cointreau et al., 1984). Ogwueleka (2009), reported urban waste generation in Nigeria in the range of 12,000 to 255,556 tons per month (Fig.1); with Lagos, the commercial nerve centre of the country, generating the highest.

Focusing on Nigeria, waste management challenges in Port- Harcourt city, used as a case study in this project, is not different from any other major cities in developing countries (Zurbrugg, 2003; Imam et al., 2008; Ugwueleka, 2009;). As part of proactive measures to preserve the environment and protect the inhabitants from hazardous wastes, the Nigerian government and several states therein have established various governmental authorities and agencies that would ensure efficient and effective mode of waste management. The following are a list of solid waste management stakeholders and major actors at both the Federal and State levels:

- National Environmental Standards and Regulations Enforcement Agency
- Federal Ministry of Environment
- State Ministries of Environment
- State Environmental Protection Agencies

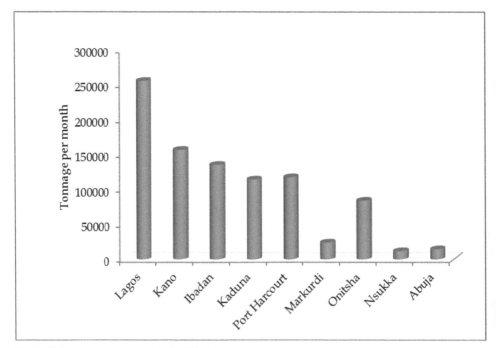

Fig. 1. Urban waste generation in Nigeria

In addition, as highlighted by ELRI (2009); various statutory regulations guiding solid waste management in Nigeria include:

- The Federal Environmental Protection Act of 1988,
- Environmental Impact Assessment Act, 1992. The purpose of the EIA Act is to, among other things, establish the magnitude of impact and mitigation measures before a decision taken by any person, authority corporate body or unincorporated body including the Government of the Federation, State or Local Government intending to undertake or authorize the undertaking of any activity that may likely or to a significant extent affect the environment. Such activities include the disposal of solid wastes in the environment,
- The National Environmental Standards and Regulations Enforcement Agency Act 2007 (NESREA ACT), which became the major statutory regulation or instrument guiding environmental matters in Nigeria. It specially makes provision for solid waste management and its administration and prescribes sanction for offences or acts which run contrary to proper and adequate waste disposal procedures and practices.
- National Environmental (Sanitation and Waste Control) Regulations, 2009. This regulation, among other things, makes adequate provisions for waste control and environmental sanitation including punishments in cases of malfeasance,
- The Harmful Waste (Special Criminal Provisions) Act prohibits the carrying, depositing and dumping of harmful waste on any land, territorial waters, contagious zone, exclusive economic zone of Nigeria or its inland water ways and prescribes severe penalties for any person found guilty of any crime relating thereto,
- The National Environmental Protection (NEP) (Pollution Abatement in Industries and Facilities Generating Waste) Regulations on the release of toxic substances and stipulates monitoring of pollution to ensure permissible limits are not exceeded; unusual and accidental discharges; contingency plans; generator's liabilities; strategies of waste reduction and safety for workers,
- The Management of Solid Hazardous Wastes Regulations. These regulate the collection, treatment and disposal of solid and hazardous wastes for municipal and industrial sources and give the comprehensive list of chemicals and chemical wastes by toxicity categories,
- The National Oil Spill Detection and Response agency, 2005 (NOSDRA ACT). This statutory regulation makes adequate regulations on wastes emanating from oil production and exploration and its potential consequences to the environment,
- The National Effluents Limitations Regulation. This instrument makes it mandatory that industrial facilities install anti-pollution equipment, make provision for further effluent treatment, prescribe maximum limit of effluent parameters allowed for discharge, and spell out penalties for contravention,
- The National Guidelines and Standards for Environmental Pollution Control in Nigeria. This was launched on March 12th 1991 and represents the basic instrument for monitoring and controlling industrial and urban pollution.

The highlighted regulations reveal that the country has the policy platform to tackle the challenges of waste management; however, waste material disposal/management in the country is still a burning issue to environmentalists. The situation is similar to that in other developing countries. The solution to waste material management or simply put, effective waste material disposal in developing countries will not be achieved if the fundamental issues regarding waste management are not aggressively addressed.

1.2 Justification

Despite the numerous existing laws and regulations, waste material disposal and management is still a great challenge facing environmental protection agencies in developing countries. This is not just a challenge to environmental sustainability but a social handicap. It is obvious that developing countries are still battling with the fundamentals of waste management. In contrast, waste management in many industrialized countries such as Germany, United Kingdom, and United States of America is governed by strict environmental guidelines and enforcement. There is no doubt that improved technology is an advantage for the developed countries but the pertinent question is: would developing countries fare better if high-tech facilities are donated to them right now? For a successful implementation of a given technology, rules and regulation must be implemented.

Using Nigeria as an example of the developing countries, a wide variety of approaches has been strategized in an attempt to tackle this environmental challenge. Some of these efforts include:

- workshops for stakeholders and academic conferences, which often end up at the level of a communiqué and or published articles, without practical impact,
- waste collection by municipal or related waste management authorities in the urban areas. In this case, residents lump different types of waste materials (food wastes, plastics, nylons, electronics, papers, glass, metal scraps etc) together at designated open dumps in the neighbourhood from where trucks convey them to an approved waste dumpsites and
- attempts by some state governments to delve into waste to wealth initiatives, involving Public-Private Partnership (PPP).

These efforts have, however, not yet solved the environmental problems posed by waste materials. The last approach is deemed to be the most viable option due to accruing benefits of dynamism, access to finance, knowledge of technologies, managerial efficiency and entrepreneurial spirit. It is, therefore, time to take a closer look and identify the rudimental causative factors to inappropriate waste management in developing countries. This is a step in the direction to finding a viable solution to this social menace.

Until recently, Port Harcourt was known as the "garden city of Nigeria" because of its neatness and the overwhelming presence of parks and gardens all over the metropolis. However, according to Ayotamuno et al., (2004), the presence of piles of refuse dotting the entire city has turned Port Harcourt to a "garbage city". The situation is such that traffic flow is often obstructed, while there is likelihood that leachates from such dumps, after mixing with rain water, have the potential to contaminate ground water (Ayotamuno et al., 2004; Adekunle et al., 2007). It is believed that PPP in waste management strategy will become more efficient if there is a concerted effort in mass mobilization/re-orientation to positively change the attitude of citizens in waste material handling and disposal. Organic waste materials can be effectively treated and disposed for land applications in the fields of environment and agriculture, paving way for a sustainable Public-Private-Partnership in solid waste management in most developing countries and Nigeria in particular (Adekunle et al., 2011).

1.3 Objectives

This chapter assesses (i) the level of population awareness, attitude and willingness to participate in value-added waste disposal for environmental sustainability in developing countries, using Port Harcourt city, Nigeria as a case study and (ii) possible factors contributing to lack of willingness to participate in value-added waste disposal.

2. Materials and methods

2.1 Study area

The Federal Republic of Nigeria comprises 36 states and the Federal Capital Territory, Abuja. Port Harcourt (Fig.2), is the capital of Rivers state, one of the 36 states, with geographical coordinates of 4⁰ 27′ 2″ North and 6⁰ 59′ 55″ East. Port Harcourt features a tropical monsoon climate with lengthy and heavy rainy seasons and very short dry seasons. Only the months of December and January truly qualify as dry season months. The harmattan, which climatically influences many cities in West Africa, is less pronounced in Port Harcourt. Port Harcourt's heaviest precipitation occurs during September with an average of 370 mm of rain. December on average is the driest month of the year; with an average rainfall of 20 mm. Temperature is relatively constant, showing little variation throughout the course of the year. Average temperatures are typically between 25 - 28°C (Wikipedia, 2011).

Port Harcourt is one of the major cities in Nigeria and the second largest port in the country after Lagos. It is a major industrial centre as it has a large number of multinational firms as well as other industrial concerns, particularly business related to the petroleum industry. It is the chief oil-refining city in Nigeria. Due to its economic vibrancy and population growth, waste material handling/disposal becomes an important environmental sustainability factor.

Fig. 2. Map of Nigeria showing Port Harcourt city

Qualitative research survey with the use of structured questionnaire was employed for this work. The respondents comprised of seven hundred, randomly selected, adults within the age range of 18 to 60 years; spread across the highly and non- highly educated, low to high income earners, males and females. All respondents lived in Port Harcourt at the time of study.

2.2 Research method

The questionnaire consisted of 30 research questions, which were grouped under 6 segments. An overview of the research question is presented as follows:

i.   socio-demographic characteristics of participants – age, gender, marital status, educational level and household size
ii. socio-economic characteristics of participants - employment status, income level, housing tenure, housing type and period of tenement
iii. awareness and participation in general solid waste management - knowledge and awareness on solid waste management, source of waste management information, waste handling method, type of waste sorting, type of waste collection system, time of waste collection and collection frequency
iv. awareness and participation in organic solid waste management - knowledge and awareness on organic solid waste management, source of information, organic solid waste management style, composting as organic solid management approach, type of composting practiced, knowledge and awareness on environmental impact of waste materials
v.   financial implication of waste management to the respondents – consideration of waste materials as raw materials, ever been paid money for waste materials generated for transformation to a beneficial end use, ever paid money to municipal authority for waste disposal, possibility of being motivated if paid some token for effective waste disposal? Disposition if asked to pay for waste disposal
vi. Willingness to participate in community based composting for value-added waste treatment and disposal – willingness to participate in community based composting, willingness to sort wastes at source for easy waste management and disposal, consideration of effective waste disposal as a necessity for environmental sustainability?

2.3 Data analysis

Primary data collected from the 700 respondents were grouped according to the variables. Frequencies were computed to obtain counts on variables' values, which were translated to percentage values. Results were then presented as bar-charts, pie-charts and Tables.

3. Results

3.1 Socio-demographic characteristics of respondents

Results on the socio-demographic characteristics of the 700 respondents (Fig.3) showed that 9% were within 18 to 21 years, 81% fell within 22 to 50 years of age; and 10% were above 50 years. The total study population (60% males and 40% females), comprised largely (70%) of married people while singles made up the 28% and the minority (2%) were either divorced (1%) or widowed (1%). Results, presented in Fig.4, showed that the majority (70%) of the respondents was educated, with a minimum of first degree in diverse disciplines; 25% had a minimum of secondary school certificate or the equivalent and only 4% were primary school graduates. Household sizes varied between 1 and 10 people. Specifically, 36% of the respondents had household size of 3 to 4 persons, 35% had 1 to 2 persons, 19% had 5 to 7 persons and 5% had more than 10 persons constituting the house.

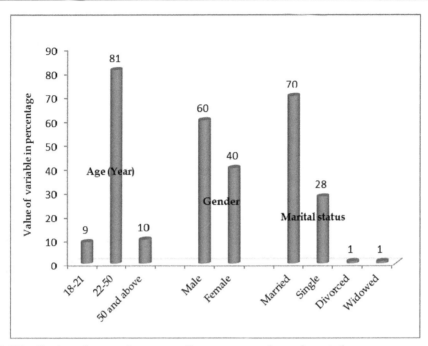

Fig. 3. Distribution of respondents according to age, gender and marital status

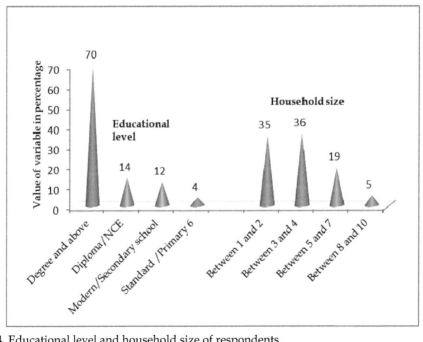

Fig. 4. Educational level and household size of respondents

3.2 Socio-economic characteristics of respondents

Results on socio-economic characteristics of the respondents are presented under the following variables: (i) employment status, (ii) income level, (iii) housing tenure, (iv) housing type and (v) period of tenement. Employment status and income level of respondents are presented in Fig.5 while information on housing regiment of respondents showing housing tenure, type and period of tenement are given in Fig.6. More than half (59%) of the participants were gainfully employed by either the Government or private sector. Only 22% were self employed while 11% were unemployed. However, 8% of them were engaged in jobs that could not be categorized under the previously mentioned job groups. Fifty-seven percent of the population perceived themselves as medium income earners as against 35% and 8% that claimed to be low income and high income earners respectively. The trend in income status was identified in the order: medium income earners > low income earners > high income earners.

Regarding housing regiment (Fig.6), 66% of the respondents were tenants, living in rented apartments and 34% of the total population lived in their own houses; 86% lived in residential areas, 14% dwelt in non-residential (4% for institutional and 10% for commercial). Period of tenement in their houses varied between 1 and 10 years for 48% of the people, less than 1 year for 29% and between 11 and 20 years for just 7% of the participants. However, 16% could not recall how long they had resided in their houses.

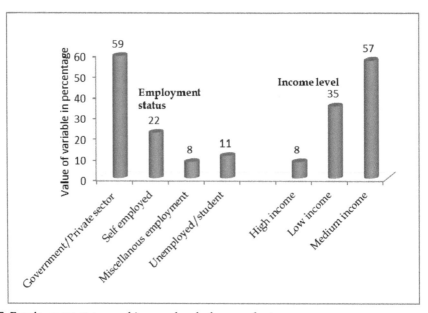

Fig. 5. Employment status and income level of respondents

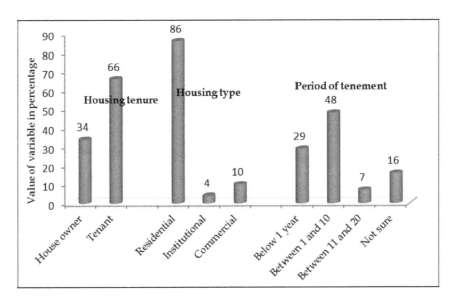

Fig. 6. Information on housing regiment of respondents showing housing tenure, type and period of tenement

3.3 Awareness and participation in general waste management

Results on awareness and participation in general waste management are presented under the following concepts (i) knowledge and awareness on waste management, (ii) sources of waste management information, (iii) waste material handling (iv) type of waste sorting: into recyclables, non-recyclables and biodegradables (v) type of waste collection systems, (vi) time of waste collection and (vii) frequency of collection. Data showed that 76% of the target population was very much aware of the negative impacts of improper waste material disposal on the environment (Fig.7).

Results, presented in Fig.8, showed that up to 85% of the people were aware of the theoretical concept of general solid waste management. Major sources of information were identified as (i) media, (ii) school and (iii) neighborhood for 46%, 13% and 4% of the respondents, respectively. Some participants (3%) claimed to have been informed by friends, 3% attributed their knowledge to other sources of information such as offices and internet facilities while up to 20% of the participants claimed ignorance. Regarding information on waste material handling at source (Fig.9), results showed that no participant really sorted waste materials at source; rather, an overwhelming majority (81%) mixed or mingled their wastes together in a given waste bin and 19% practiced indiscriminate disposal. By this, they disposed of their waste materials at convenience, without discretion on environmental impact of such an action. This, cumulatively, gave a total of 100% of the study population involved in non-segregation of waste at source.

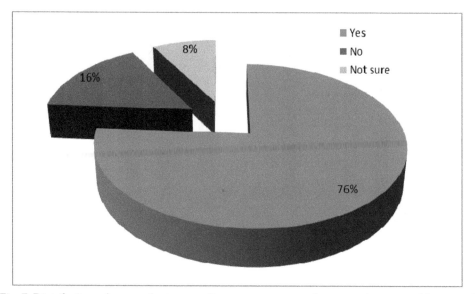

Fig. 7. Distribution of respondents regarding awareness on adverse environmental impact of improper disposal of waste materials

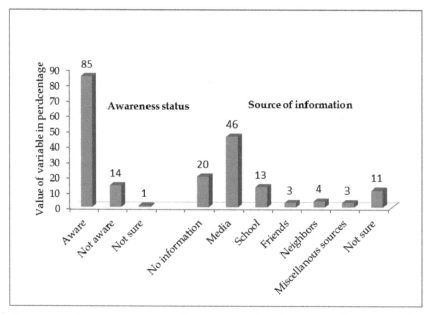

Fig. 8. Awareness status on general waste management and corresponding source of information

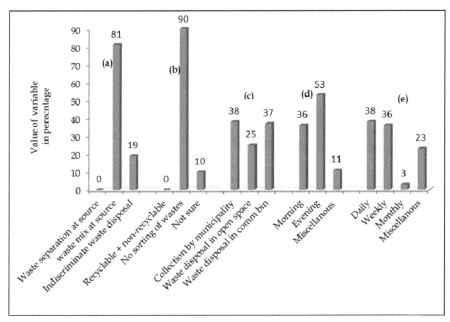

Fig. 9. Waste disposal strategy via handling (a), sorting (b), waste collection technique (c)
time of disposal (d) and frequency of disposal (e)

Sequel to perceived lack of proper understanding on the concept by the respondents and
benefits of waste segregation at source, a more detailed probe carried out on this subject
revealed that none of the participants separated their waste materials into recyclables and
non-recyclables materials at source. With this, 90% of the populace confirmed that they did
not practice waste separation at source. Ten percent of the people could not categorize their
method but they were definite that their method did not fall under waste separation at
source. Figure 9 showed that collection of waste materials for disposal by municipal
authority was reported by 38% of the participants while waste material disposal in open
space (illegal waste dumpsite) was acknowledged by 25% of them and 37% of the
participants disposed their wastes in community bins.

3.4 Awareness and participation in organic waste management

The results on the awareness and participation in organic waste management, shown in
Fig.10, revealed that 67% of the study population acknowledged theoretical information on
organic solid waste management. The remaining 33% were not very knowledgeable on this
subject matter. Source of information decreased in the following order: media (44%) >
school (13%) > information from friends (3%) > information from the neighborhood (1%).
Thirty-nine percent of the participants could not trace their source of information. On
organic waste management strategy, 92% of the populace disposed wastes into the
collection bin, while the minority (5%) disposed biodegradables into the garden/farm

(Fig.10). None of the respondents practiced composting as effective disposal method for biodegradable organic waste materials.

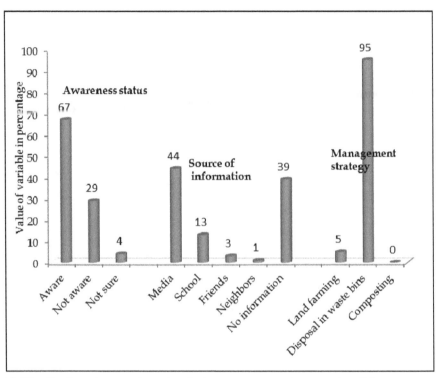

Fig. 10. Distribution of respondents in relation to awareness on organic solid management, source of information and management strategy

3.5 Knowledge and awareness of composting technology as a value-added disposal technique for organic wastes

Results on the survey of theoretical knowledge of the respondents on composting techniques (Fig.11) showed that 40% of the study population was not aware of composting technology. Data showed that the most common composting technology known to respondents was heap composting as acknowledged by 19%, against on-farm composting, windrows, trench/pit, static in-vessel and mechanized/automated composting methods acknowledged by 10%, 14%, 6%, 3% and 3% of the study population, respectively. After sensitizing the respondents on composting technology as a value-added organic waste material disposal technique, there was a shift in perception. Data showed that 82% of the total study population was willing to participate in community based composting for environmental sustainability. Moreover, 84% of them became eager and willing to practice waste segregation at source to facilitate community based composting (Fig.12).

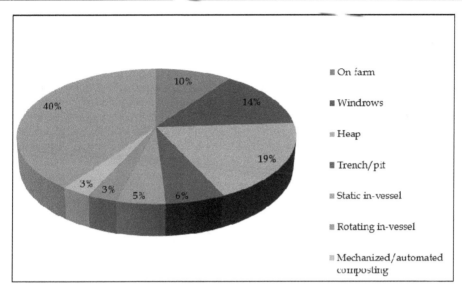

Fig. 11. Information on theoretical knowledge of composting techniques

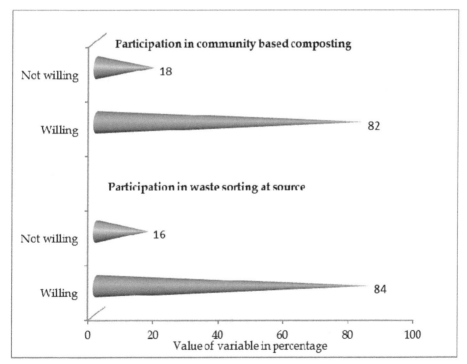

Fig. 12. Distribution of respondents in relation to willingness to participate in waste sorting
at source and community base composting for value-added disposal of organic waste
materials

3.6 Financial implication of waste management

The response of participants on financial implication of waste management is presented in Table 1. Exactly 84% of the populace considered waste materials as potential raw materials that could procure financial empowerment, but only 69% had actually derived monetary benefits from waste materials. For effective waste material disposal, 58% of the respondents have at one time or the other paid municipal waste management authority. Again, 86% of the people would be motivated to manage their wastes better if they could derive some degree of financial boost from their waste material and 79% would still be motivated even if asked to pay some specific amount of money to achieve effective disposal of their waste materials. The reason is not farfetched as 94% of them were convinced that adequate waste material management is vital to environmental sustainability (Fig.13).

S/N	Inquiry	Response (%)		
		Yes	No	Not sure
1.	Ever considered waste materials as raw materials?	84	13	3
2.	Ever been paid money for your waste materials?	31	69	Nil
3.	Ever paid to municipality for waste material disposal?	58	39	3
4.	Will you be motivated if paid some token for effective disposal of your own wastes materials?	86	10	4
5.	Will you still be motivated if asked to pay some token for effective disposal of your own waste materials?	79	17	4

Table 1. Response of participants on financial implication of waste management

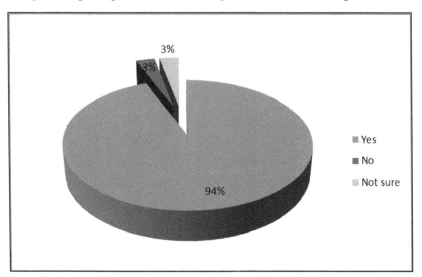

Fig. 13. Distribution of respondents regarding the consideration of waste management as an important factor to environmental sustainability

3.7 Possible militating factors against waste material sorting at source and community based composting

Results on the evaluation of potential factors influencing waste material sorting at source and community based composting are presented in Table 2. Regarding, waste sorting at source (WSAS), the impact of some suspected militating factors were obtained in the range of 15 - 36% for age, 17 - 20% for gender, 15 - 60% for marital status, 12 – 57% for educational level, 12 - 25% for income status, 17 - 20% for housing tenure, 14 – 42% for housing type, 15 - 19% for period of tenement, and 12 – 24% for household size.

S/N	Possible influencing factor	Total number of respondents in response to a given factor	Fraction of population opposed to waste sorting at source		Fraction of population opposed to community based composting	
			Actual number	*Impact (%)	Actual number	**Impact (%)
1.	Age (Years)					
	18-21	64	23	36	16	25
	22-50	567	83	15	88	16
	≥ 50	69	21	30	18	26
2.	Gender					
	Male	420	71	17	71	17
	Female	280	56	20	51	18
3.	Marital status					
	Married	489	73	15	63	13
	Single	197	46	23	55	28
	Divorced	5	3	60	1	20
	Widowed	9	4	44	3	33
4.	Educational level					
	Standard 6/Primary school	28	16	57	11	39
	Modern school/secondary school	87	29	33	23	26
	Diploma/NCE	95	19	20	9	9
	First degree and above	490	60	12	79	16
5.	Income level					
	Low	240	60	25	43	18
	Medium	402	50	12	55	14
	High	58	11	19	13	22
6.	Housing tenure					
	Owner	240	48	20	44	18
	Tenant	460	77	17	78	17

7.	Housing type					
	Residential	598	85	14	97	16
	Institutional	30	10	42	5	17
	Commercial	72	30	42	20	28
8.	Period of tenement (Year)					
	Less than 1	206	39	19	40	19
	Between 1 and 10	334	52	16	78	17
	Between 11 and 20	46	7	15	4	8
9.	Household size (people)					
	Between 1 and 2	243	39	16	51	21
	Between 3 and 4	252	38	15	40	16
	Between 5 and 7	133	31	23	22	17
	Between 8 and 10	37	9	24	7	19
	Ten and above	33	4	12	2	6

*Computed as the number of population opposed to WSAS relative to the total number of study population in response to a given factor, ** computed as the number of population opposed to community based composting relative to the total number of study population in response to a given factor.

Table 2: Evaluation of possible influencing factors to unwillingness in the participation of waste material sorting at source and community based composting

In the perspective of community based composting, the contributions of potential militating factors varied from 16 -26% for age, 17 -18% for gender, 13 - 33% for marital status, 9 - 39% for educational level, 14 – 22% for income level, 17 -18% for housing tenure, 16 -28% for housing type, 8 -19% for period of tenement and 6 -21% for household size.

4. Discussion

This study revealed that most of the respondents were very ignorant of best practices in waste material disposal and management. About 84% of them considered waste materials as potential raw materials but were incapacitated by lack of knowledge. The financial benefit from waste materials as acknowledged by 69% of the respondents came from their own ingenuity. They were involved in practices such as trading in old clothes, shoes, boxes, plastic bottles and glass bottles to generate income. It was also revealed from this survey that the impact of population awareness on waste disposal and management from the media (46%) and schools (13%) were very poor. The factor commonly considered as basic to inappropriate disposal of wastes in developing countries is poverty. Some environmentalists are of the opinion that improper waste material handling in developing countries is compounded by the vicious cycle of poverty; population explosion and decreasing standard of living (Zerock, 2003; Al-Khahb et al., 2007; Ogwueleka, 2009). Surprisingly, the findings from this study did not tow the same direction. This study revealed that impact of income status on either waste segregation at source or willingness to participate in value added waste management, typified by community based composting, was below 30%; varying from 12 to 22%. Similarly, household population impact ranged from 6 to 24%, indicating that poverty and population are not really the foundational causative factors to unsafe waste material disposal and management in developing countries. Pre-survey analysis revealed zero practice of waste segregation at source but proper enlightenment of the respondent awakened their interest to participate in segregation by 84% and value-added waste management (community based composting) by 82%. It then confirms that positive attitudinal change in the citizens towards best solid waste management practices will be achieved with proper education and mobilization of the populace. Non-segregation of waste materials at source due to ignorance, resulting in lack of easy access to raw materials to work with is, therefore, a major mitigating factor to the Public-Private-Partnership enterprise in solid waste management. Consequently, the investors largely depend on human scavengers who source for raw materials by handpicking waste materials lumped together in the open dumpsites, as shown in Fig. 14.

Fig. 14. Typical informal sector involved in waste recycling in developing countries

This practice is a messy and unhealthy procedure and also undermines the efficiency of Public-Private Partnership. Effective waste material disposal is paramount to a successful Public-Private Partnership. Despite policy promulgation by government authorities and workshops at high places, if the citizens are not properly mobilized or carried along in waste disposal/management program, effective waste disposal in developing countries will still be a mirage.

The respondents were not unaware of the adverse impact of improper waste handling and disposal, prevalent in their environment. These include unsightly aesthetics and corresponding reduced tourist appeal, air pollution, surface water contamination, land pollution, drainage blockages resulting in flooding and water borne diseases. They also acknowledged that the current solid waste handling, pose risk to public health and the ecosystem in general. They appear to be helpless with the unfavorable and non-conducive conditions of waste management. For instance, there is no proper infrastructure in most cities where one can conveniently dispose segregated wastes. Even if convinced to segregate at household levels, where will these wastes be finally conveyed? They will ultimately end up in a common community bin or dumpsite! Regardless of the form of waste management strategy adopted by municipal authorities, the foundation to value-added waste material transaction is linked to proper handling of the waste materials at source. The problem of waste material management in the developing countries, when properly investigated will be traced to faulty foundations.

The effective disposal of municipal waste material is a major responsibility of the state and local government agencies that often attribute failure to effectively manage waste materials to the fact that the capacity of most municipal services is overwhelmed by rapid, unplanned for population growth against existing infrastructure. The current study revealed that waste material handling at source, as acknowledged by the study population, fell below internationally accepted best management practices in waste management. Results from this study suggest that these agencies are still battling with waste management as a result of less attention paid to the basics of effective waste disposal. Adequate waste material management is beyond a technical issue. It no doubt involves institutional, social, legal and financial aspects but beyond these, involvement of stakeholders - the populace, is paramount. When successful waste material disposal becomes a priority, the mass/citizens have to be carried along by pragmatic and grassroot awareness and enlightenment campaigns. Figure 15 illustrates the current trend in solid waste management/disposal in developing countries as opposed to the ideal focus for value-added waste management approach (Fig.16).

In addition, an important factor that will promote the success of privatization of solid waste management in the developing countries is the enforcement of waste material segregation at source. If the different kinds of waste materials are lumped or mixed together in one collection vessel, which could range from paper bag, polyethene bags, plastic buckets to metallic containers; the values of the materials are lost to a large degree so also the viability of Public-Private-Partnership. For PPPs to succeed, the raw materials procured from wastes must be in good shape. From this study, the major reason solid waste management in developing countries is a herculean task is not because the quantity and complexity of the materials are greater than the ones generated in the developed nations, neither is it essentially due to lack of high - tech facilities. The reason why privatization of solid waste management is not yet a thriving business in developing countries is not due to over population or increased trend in rural-urban migration. There is no doubt to obvious challenges to waste management in developing countries which include:

- material composition, density and quantity
- inefficient waste collection method
- access to waste materials and improper disposal
- bottlenecks from political and economic frameworks
- technological platform (facilities and human expertise).

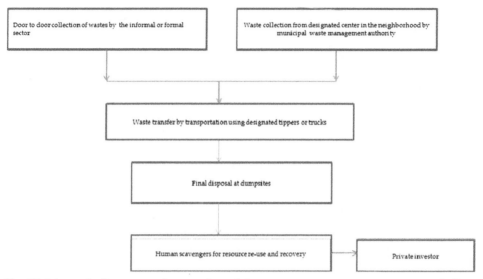

Fig. 15. Schematic diagram on the overview of the current waste management system in Port Harcourt and Nigeria in general

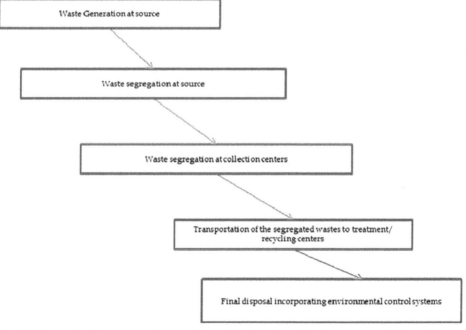

Fig. 16. Schematic diagram on the overview of ideal value-added waste material handling and disposal

These factors notwithstanding, this study has shown that waste management in developing countries could take a leap in the direction of Public-Private-Partnership if the citizens are properly oriented. Ignorance and lack of energetic mobilization of the populace are primary factors to low sanitation habit of most people in developing countries. It is often asserted that people living in developing countries are chronically of low hygiene but let us pause and ponder. When people from developing countries travel and live in developed countries, do they not conform to the laws and regulations on waste disposal? This is because of the awareness they get in such places. Furthermore, are all of the immigrants from the developing countries to the developed countries educated and rich? It will, therefore, not be very correct to say that people in developing countries enjoy the poor sanitary condition they find themselves in. In developing countries, an attitudinal change is expedient to effective waste materials disposal and management for a healthy environment and this can only be achieved through proper orientation and mobilization of the populace.

After enlightenment of the respondent, the fraction (84%) that are willing to participate in waste segregation at source and 82% in community based composting points to the eagerness of the citizens to see their waste materials transformed to resource materials that can be utilized beneficially. This is an indication that they consider source segregation as a step in waste management for environmental sustainability. It then follows that if properly informed, mobilized and conditions made favorable, the citizens will actively participate in effective waste disposal and value-added solid waste management such as segregation at source and community based composting scheme, for a cleaner and healthier environment. This step will also facilitate the Public-Private-Partnership in waste management.

Improper disposal of waste materials constitutes environmental and health hazards and contravenes an aspect of the Millennium Development Goals (MDGs), in particular, Goal 7 that focuses on environmental sustainability. Proper and effective waste disposal is an important component of environmental sanitation and sustainability. This makes waste segregation at source a practical and mandatory measure to attaining the goal 7 of the MDGs, which can only be achieved through proper information dissemination to the citizenry. This survey revealed positive attitudinal change as an area that should be worked on by relevant waste management authorities in developing countries for environmental sustainability. This study also showed that information performance on this subject decreased as follows: media (46%)> school (13%)> neighborhood (4%) > miscellaneous (internet, office: 3%). How different it would be if the percentage impact from school on information dissemination was very much higher (on the average of 70%)! This is an avenue where the citizens in developing countries could be captured young and proper waste handling becomes part and parcel of their daily activities. This is where the hearts and mind could be positively influenced and piloted towards effective and value-added waste disposal and management.

4.1 Role of population perception on value added waste material disposal

Perception is a way of regarding, understanding or interpreting something; a mental impression of a given phenomenon, in this case, solid waste management. More emphatically, population perception of waste management describes the whole process of how the populace comes to know what is going on regarding best practices in waste management. Simply put, it is the emotional and mental disposition of the mass/citizens towards waste management protocols. Awareness and enlightenment programs through information, education (formal and informal), capacity building, coupled with implementation and execution of laws and regulations on proper waste disposal will affect the population's perception and willingness to participate in best waste management practices. Presently, municipal governments in Nigeria, particularly Port Harcourt city, are solely responsible for all waste related activities. The possibility that some personnel involved in waste management under the auspices local authorities do not know better should not be ruled out, hence, it is also important to get them well informed on effective waste material disposal strategies. Population participation in waste management protocols is the process by which individuals in the families and communities; both old and young understand and take responsibility for their generated wastes. The cooperation of the citizens leads to a successful solid waste management system in any city. The people should be involved in the following processes:

- proper waste storage at source
- waste segregation at source
- waste collection and transfer to an approved point in the neighborhood where wastes are also segregated prior to collection by municipal authority.

This organized waste handling results in safe and environmentally friendly waste material disposal. There is no denial about some ongoing orientation activities and programs in developing countries, Nigeria inclusive, on waste management but the clarion call is that a more pragmatic approach should be adopted. Result oriented, goal achieving strategies are needful. Mass mobilization in ways and languages the people understand should be adopted. In Nigeria, the most current program is an activity that requires residents to spend the last or first Saturday morning of each month to clean their surrounding and clear waste materials dumped on the streets. This implies that in most places, improperly disposed wastes pile up, waiting for the appointed day of state sanitation exercise. This type of program provides a temporary measure, ensuring that proper sanitation is carried out only twelve times a year in each city.

Benefits of effective population participation in waste management include:

- reduction of waste generation at source
- increased re-use of materials before final discarding
- safe collection of wastes
- proper storage of organic biodegradable waste for value-added disposal
- adequate storage of recyclable waste materials separately at source for value-added disposal
- reduced indiscriminate discarding of wastes, which minimizes the quantity of litters on the street and those thrown into drainages
- provision of less polluted raw materials for private investors, in contrast to the more polluted waste materials picked from the waste dumps as currently practiced
- improved environmental aesthetics
- improved public health
- viable Public-Private-Partnership in waste management

4.2 Approaches to enhancing population participation in value-added waste material disposal and management

The family and community are pivotal to all human activities including waste material disposal. Avoiding or ignoring these important stakeholders creates a void in waste management system. Public participation is an important factor to the success of safe waste material disposal and as a matter of necessity must be encouraged. In most cases, countries in developing world comprise of diverse ethnic groups, cultures and beliefs, so it is essential that each country/city decides on the methodology that is best suitable to the citizenry, seeking their cooperation and effective participation. However, basic strategies to enhancing population perception and participation in best practices in waste management are briefly discussed:

4.2.1 Identification of stakeholders

Identification of stakeholders involves assessing the group of people that constitute the populace in a community. The stakeholders can be categorized in a variety of ways, using different criteria such as age, gender, education, income level and marital status. They can also be grouped according to housing schemes such as those living in the residential areas, commercial areas and institutions. Dividing the populace into these fragments gives a more effective outreach. For instance, addressing the highly educated and the illiterates using the same method will obviously bring disparity in understanding. Waste management seminars, workshops, conferences and publications will be of little use to the uneducated. The point made here is that there should be an all encompassing outreach to the grass root. This creates an enhanced interaction with the populace, which is the most important issue because of the diversity in the level of awareness and sensitivity in each group.

4.2.2 Community mobilization

Reaching the community is quite a difficult task but an effective methodology that could be helpful. To achieve the desired result in this area, the local authority in waste management could select reliable community representative that could initiate consultative processes. This will facilitate the understanding of the population perception and expectation, which will in turn help the authority on how to tackle the people's attitude, influence their willingness to participate in safe waste material disposal and the relevant option in effective value-added waste management. It is important to stress that the community representative are more effective if they are able to speak the language of the people for enhanced interaction.

4.2.3 Public awareness campaigns

This is a process in which people are made aware of the problems of improper waste material disposal and the benefits of safe disposal in waste management. The essence of public awareness, public education and motivation programmes is to (i) give diverse possibilities for public/stakeholder interaction, (ii) influence public perception (iii) induce attitudinal change, (iv) clarify doubts and negative perceptions (v) give constructive follow-up and (vi) provide platform for policy makers.

The enlightenment campaign should be conducted in the most suitable mode that appeals to the populace in a given region. These activities will ensure that people become aware of the problems and negative environmental impacts of waste accumulation in streets and gutters, emphasizing how it directly affects their lives. The message should promote reduced waste generation and provide information on safe handling of generated wastes. The campaigns should inform the populace on the waste management program outlined by the government and possible penalties for defaulters. Taking the issue of road safety campaign in Nigeria as an illustration; until recently, most Nigerians drove cars without fastening the seat belt but the concerted efforts of the Federal Road Safety Commission (FRSC), who embarked on grass root awareness campaign, implementation and execution of promulgated policy and dispensing punishment to defaulters; has brought a positive change to the perception of the use of seat belt in driving. Similarly, some other Institutions in Nigeria such as the Shell Petroleum Development Company (Nigeria) Limited, have performed excellently in educating their workers on safe waste material disposal. If significant levels of success have been achieved by the two cited examples, it indicates that adequate awareness and enlightenment campaign on safe waste materials disposal will achieve a great level of success. Public awareness could be actualized through the under listed approaches:

Environmental education: this involves group meetings in the community, workshops, exhibitions, seminars, lectures series, panel discussions, dramas, posters, banners and fliers. Institution of waste management clubs in primary, secondary and tertiary schools; propagation of slogans that promote clean environment and outreach services/programs through the media in both national and local languages should be encouraged.

Open education: in most practical cases, it may not be possible to reach out to the relevant stake holders through the conventional educational institutions. Mass/open education through the formal and informal sector therefore comes to play. The vital factor here is effective communication. Avenues to achieving this include print, web/internet and audio-visual media. Also effective are the use of entertainment media such as cinema outreach, comedy forum, street plays, street dances, animations (e.g. puppet shows and cartoons), carnivals and reality shows. The display of pamphlets, handbills, posters, banners, and fliers with good photographs and messages with few and readable words also raise awareness. In addition, messages can be conveyed by paintings on walls and buses.

Academic curricula: Inclusion of waste management for environmental sustainability in academic curricula at all levels of education will immensely contribute positively to population perception and attitudinal change. If the citizens are mobilized at a tender age, safe waste material disposal then becomes a natural habit. This promotes sense of responsibility and best practices in waste management becomes inculcated in the citizens very early in life and their hearts and minds are captured; children from homes where parents are not formally educated become powerful educators/communicators, bridging important gaps. The establishment of literary and debate clubs in waste management in school, who would be involved in various competitions, would reinforce safe practices. Prizes and awards to winning contestants, who will be publicized as role models for others would further strengthen the efforts, bringing life and vitality to issues of waste management.

Involvement of faith based organizations: the involvement of religious leaders would facilitate safe waste material disposal in developing countries where religious leaders play significant roles in influencing peoples' perception on sensitive issues. Seminars/capacity building workshops on waste management issues can be organized by government for faith based leaders, who in turn will speak to their followers in much more effective ways.

Mobilization of women: the female gender is an important change agent in most societies of the world, especially in the developing countries. "Train a woman, train a nation" says the old adage. Furthermore, women are more accountable to the maintenance of health and hygiene in most homes.

Involvement of health sectors: the medical practitioners such as doctors, nurses and paramedics are also in the position to influence the populace. They are held in high esteem by their patients, so counsels from them will be appreciated and practiced by their clients.

Involvement of non-faith/governmental based organizations: Many non-governmental organizations (NGOs) should be encouraged financially by government and international bodies to actively participate in mass mobilization and public enlightenment regarding safe waste material disposal. In most cases, NGOs have personnel equipped with good mass-communication and education programs who could develop programs for the public.

4.2.4 Enforcement of promulgated policies

The very many policies, totaling up to eleven, on environmental issues in Nigeria have not effectively addressed the problems of inadequate waste material disposal. Theses statutory laws and regulations are often reviewed, revised, updated and often leads to the promulgation of new laws thereby increasing their numbers but not necessarily their effectiveness. There are diversities of understanding potentials in a group of people. In an attempt to exhaust all avenues of educating the populace to willingly participate in safe waste material disposal for a sustainable environment, it should be noted that there are differentials in the understanding rate for a given hum population. Their norms, values, commitment and dedication to a national course also differ. It then becomes necessary to use alternative tool in bringing about attitudinal change and demonstrated change in perception. That tool is enforcement by the appropriate agency without compromising standards. In this process of enforcement, defaulters should be punished to serve as deterrent to adamant ones. The laws enforcement agent should live above boards in terms of corruption. Any officer involved in corrupt practices should be brought to book.

4.3 Value-added solid waste disposal management

The most viable waste recycling procedure currently practiced in developing countries is equally dangerous. In these countries, resource recovery and recycling activities on waste management are driven by the informal sector. Populations of low socio-economic status consisting of both the elderly and the young (scavengers) pick valuable items found in the waste dumps (Fig.14), including bottles, metallic wares, perfume containers, and sell them to private sectors. There is a school of thought that this is a means that provides livelihood to thousands of poverty stricken citizens. These human scavengers, who use their bare hands and spend long hours in the dumps, sifting through the rubbish for valuables, are exposed to health risk and when they become infected with contagious diseases such as diarrhea, dysentery amongst many, the populations of high socio-economic status are not spared, putting the nations at risk. Furthermore, lack of personal protective equipment puts the scavengers in direct contact with needles and different types of hospital wastes, exposing them to diseases such as HIV and AID (Agunwamba, 1998).

Value-added solid waste disposal focuses on the transformation of waste materials to useful end-products or raw material for possible industrial applications. This is a trend that has gained global acceptance but has not blossomed in the developing countries (Adekunle & Adekunle, 2006; Garg et al., 2007). The promotion of population participation in waste management brings about feasible and viable Public-Private - Partnership. For PPP to thrive in developing countries, the waste collection system must undergo a drastic change. The advent anchors on waste segregation at source, which is currently being neglected. The objectives of PPPs are to establish an integrated solid waste management system mode to achieve excellent level of collection and disposal, whereby the private sector will be involved in the process; from collection of waste material till its final disposal. Aside from waste segregation at source, meeting infrastructure needs is a militating factor against PPP in the developing countries. Viable options (Lavee & Khatib , 2010; Selke, 2002) in the evolution of PPP in solid waste management include:

Metal scrap recycling: the scrap metal recycling industry encompasses a wide range of metals but are divided into two basic categories: ferrous and nonferrous. Ferrous scrap is metal that contains iron. Iron and steel (which contains iron) can be processed and re-melted repeatedly to form new objects. Most common nonferrous metals are copper, brass, aluminum, zinc, magnesium, tin, nickel, and lead. Commonly recycled metals (by volume) are iron and scrap steel, copper, aluminum, lead, zinc, and stainless steel. Sources of ferrous scraps include: mill scrap (from primary processing), used construction beams, plates, pipes, tubes, wiring, old automobiles and other automotive scraps, boat scrap, railroad scrap, railcar scrap and miscellaneous scrap metals. Aluminum is the most widely-recycled nonferrous metal. The major sources of nonferrous scrap are industrial or new scrap and obsolete scrap. Industrial or new scrap may include: aluminum left over when can lids are punched out of sheets, brass from lock manufacturing, copper from tubing manufacturing. Other major sources for metal scraps are: copper cables, copper household products, copper and zinc pipes and radiators, zinc from die-cast alloys in cars, aluminum from used beverage cans, aluminum from building siding, platinum from automobile catalytic converters, gold from electronic applications, silver from used photographic film, nickel from stainless steel and lead from battery plates (OSHA, 2008).

Waste plastic recycling: this is the process of recovering scrap or waste plastics and reprocessing the materials into useful products, sometimes completely different in form from their original state. Often, this could involve melting down plastic bottles and then casting them as plastic chairs and tables. Academic materials such as rulers are also produced by plastic recycling. Typically, a plastic is not recycled into the same type of plastic, and products made from recycled plastics are often not recyclable. Plastic wastes also serve as secondary feedstock for blast-furnace coke production (Fortelny et al., 2004; Melendi et al., 2011).

Waste paper recycling: there are three major categories of paper materials that can be used as feedstock for making recycled paper: mill broke, pre-consumer waste, and post-consumer waste. Mill broke is paper trimmings and other paper scrap from the manufacture of paper, and is recycled internally in a paper mill. Pre-consumer waste is material which left the paper mill but was discarded before it was ready for consumer use and post-consumer waste is material discarded after consumer use such as old magazines, old newspapers, office paper, old telephone directories, and residential mixed paper. Paper suitable for recycling is called "scrap paper", often used to produce molded pulp packaging (Huhtala, 1997; Merrild et al., 2008).

Waste tire recycling: tires are often recycled for use on basketball courts and new shoe products. However, material recovered from waste tires, known as "crumb," is generally only a cheap "filler" material and is rarely used in high volumes. Tires can be recycled into, among other things, the hot melt asphalt, typically as crumb rubber modifier - recycled asphalt pavement. Tires can also be recycled into other tires and discarded tires that are not recycled through retreading could provide a source of hydrocarbons for use as fuel, feedstock materials. In summary, waste tire recycling practices include retreading, recycling as crumbs rubber and combustion for thermal energy (Wolsky & Gaines, 1981; Yang, 1993; Jang et al., 1998).

Waste to energy initiatives: waste-to-energy conversion involves the processing of many different types of unusable waste streams into heat, electricity, and other forms of energy. A variety of technologies are used to convert waste into energy. Several of the more prominent technologies include: (a) incineration, which possesses low efficiency and high environmental burdens; often requires a large footprint and high costs, (b) bioconversion, which has a low throughput, low conversion rate, and large footprint, and requires large volumes of water per unit mass bacteria, (c) plasma gasification which requires a large-scale system and demands high volume input for economic feasibility; and (d) downdraft gasification, which demands a small-scale system and requires low volume input and results in the cleanest syngas (Wolsky & Gaines, 1981).

Recycling of biodegradable organic wastes: biodegradable waste materials are putrescible, therefore, can undergo decomposition. The common recycling methods for these are (i) anaerobic digestion to produce biogas and (ii) composting to produce formulations useful in the fields of agriculture environment (Adeoye et al., 1994; John et al., 1996; Adekunle, 2011; Adekunle et al., 2011). Major mitigating factors to these initiatives in developing countries include low product quality, lack of guidelines for product acceptability and low market demand (Adekunle, 2010). All these factors are traceable to lack of technology and human expertise in science and technology of organic solid waste management.

Bearing in mind that the nature of wastes generated in developing countries constitutes largely of organic materials, recycling of organic solid wastes appears to be a viable waste treatment and disposal method and perhaps the most feasible due to its viability under low-tech infrastructure conditions. This study has clearly demonstrated the ignorance of the respondents regarding composting technology but it is believed that for a more efficient public-private-partnership in organic waste management strategy, there must be a concerted effort in mass mobilization/ re-orientation to positively change the attitude of citizens in waste material handling and disposal. If the perception of the citizens is steered towards proper management of biodegradable waste materials via composting technology, it will greatly enhance the reduction of the waste stream going to both legitimate and illegal

Assessment of Population Perception Impact on Value-Added Solid Waste Disposal in Developing Countries,
a Case Study of Port Harcourt City, Nigeria

189

waste dumpsites. In addition, this move will contribute positively to reduced emission of methane (CH_4), a greenhouse gas, from decomposing organics. By implication, composting technology; under best management practices, will impact positively on the environment, not only in terms of land applications but also in the abatement global warming and extreme weather events.

4.3.1 Value added waste material disposal via composting technology

Reports from literature indicate that over 50% of wastes in a typical developing country could be readily composted; being biodegradable in nature (Hoornweg et al., 2000; Ogwueleka, 2009; Adekunle et al., 2011). Composting is a cornerstone of sustainable development, which is neglected within integrated municipal solid waste management in developing countries. It is a technology that does not necessarily involve complex infrastructure. Composting is certainly not a panacea to all the waste management problems in developing countries but it is an important component within most integrated municipal waste management strategies. It can be applied to various types of wastes ranging from municipal solid wastes (biodegradable materials) to wastes generated in the oil industry such as soil, sludge and oil based mud (OBM) drill cuttings (McCosh & Getliff, 2004).

Composting is a simple process where optimization efforts are used to increase the rate of decomposition (thereby reducing costs), minimize nuisance potential, and produce a clean and readily marketable finished product. Composting helps to increase the recovery rate of recyclable materials such as paper, glass, plastics; if wastes are source separated. Essential benefits of composting as highlighted by Hoornweg et al., (1999) are;

- increases overall waste diversion from final disposal, especially since as much as 80% of the waste stream in low- and middle- income countries is compostable
- enhances recycling and incineration operations by removing organic matter from the waste stream
- produces a valuable soil amendment- integral to sustainable agriculture and remediation of contaminated soils such as crude oil impacted sites and those impacted by chlorinated hydrocarbons
- promotes environmentally sound practices, such as the reduction of methane generation at landfills
- flexible for implementation at different levels, from household efforts to large-scale centralized facilities.

The second phase of this study, a pilot scale demonstration on the utilization of indigenous biodegradable waste materials, sourced from Port Harcourt city and environs in the bioremediation of crude oil impacted environmental matrices (soil, sludge and OBM drill cuttings) via composting technology, is approaching the final stage. It is postulated that by the success of this venture, the under listed benefits will be achieved:

- waste to wealth initiative
- community based organic solid waste management
- human capacity building
- job creation for the unemployed youths
- poverty alleviation
- improved environmental sanitation through safe waste material disposal

5. Conclusions and recommendations

Conclusions reached from this study are:

- over 70% of the study population were aware of the negative impact of unsafe waste material disposal
- the current public awareness and mobilization on safe waste material handling, treatment and disposal is very low (below 50%)
- waste segregation at source is rarely practiced
- waste mix at source is currently practiced by virtually all respondents
- impact of municipal authority in effective waste collection is below 50%
- the people were willing to participate in waste segregation at source and composting technology as value- added waste material disposal, if properly educated and
- population awareness and mobilization is pivotal to value-added waste disposal in developing countries.

It is recommended that machinery to enhance positive population perception on value-added waste material disposal be put in place. Basic tools in the machinery must constitute aggressive grass root orientation programs, environmental education and mobilization and enlightenment campaigns. Further studies should focus on (i) pilot scale trials on waste segregation at source using selected communities in urban areas of the developing countries (ii) community based composting schemes aimed at formulating composts suitable for land applications and bioremediation purposes.

6. Acknowledgements

The contributions of the following people: Uche Akosa, Uloma Annan, Ikubiesika Adolphus-Stanley and Ebinum Oghenekaro of Remediation Department, SPDC, in the area of questionnaire administration are acknowledged. Also acknowledged is the financial and technical support of the Remediation Department of The Shell Petroleum Development Company, Port Harcourt, Nigeria.

7. References

Adekunle, A.A & Adekunle, I.M. (2006). Creating awareness on solid waste re-use as organic fertilizer in Nigeria. Imobvare, E. (Ed). *Proceedings of the 4th Annual National Conference of the Senate on the Nigerian Environment*, held at Port Harcourt, Nigeria, July 5 - 7, pp. 126 –132.

Adekunle, I.M (2011). Bioremediation of soils contaminated with Nigerian petroleum products using composted municipal wastes. *Bioremediation Journal*, 15 (4): 1-13. Doi:10.1080/10889868.2011.624137.

Adekunle, I.M., Adekunle, A.A., Akintokun, A.K., Akintokun, P & Arowolo,T.A (2011). Recycling of organic wastes through composting for land applications: a Nigerian experience. Waste Management and Research, 29(6):582-593. Doi: 10.1177/0734242X10387312. Publisher: International Waste Management Association, Netherlands.

Adekunle, I.M., Adetunji, M.T., Gbadebo, A.M. & Banjoko, O (2007). Assessment of groundwater quality in a typical rural settlement of south west Nigeria. *Int. J. Environ. Res. Public Health,* 4(4): 307 -318.

Adekunle, I.M. (2010). Evaluating environmental impact from utilization of bulk composted wastes of Nigerian origin using laboratory extraction test. *Environmental Engineering and Management Journal,* 9 (5): 721 -729.

Adeoye, G.O., Sridhar, M.K.C & Mohammed, E.O (1994).Poultry Waste Management for Crop Production:

Nigerian Experience. *Waste Management & Research,* 12 (2) 165 -172.

Agunwamba, J.C., Egbuniwe, N & Ogwueleka, T.C., (2003). Least cost management of solid waste collection. *Journal of Solid Waste Technology and Management,* 29 (3): 154-167.

Agunwamba, J. C., (1998). Analysis of scavengers' activities and recycling in some cities of Nigeria. *Environmental Management,* 32 (1): 116-127.

Al-khatib, I.A., Arafat, H.A., Basher, T., Shawahneh, H., Salahat, A., Eid, J & Ali, W (2007). Trends and problems of solid waste management in developing countries: A case study in seven Palestinian districts. *Waste Management* 27: 1910–1919

Ayotamuno, J.M. & Gobo, A. E (2004). Municipal solid waste management in Port Harcourt, Nigeria: Obstacles and prospects. *Journal of Environmental Quality: An International Journal,* 15 (4): 389 -398.

Baker T.E. (2003). Evaluation of the Use of Scrap Tires in Transportation Related Applications in the State of Washington. Available from

http://www.wsdot.wa.gov/biz/mats/Draft%20Final%20Report%20Version%205.pdf

Cointreau, S. J., (1982). Environmental management of urban solid waste in developing countries: a project guide. Urban Development Technical paper No 5. The World Bank, Washington, DC. June .

Cointreau, S.J., Gunnerson, C.G., Huls, J. M. & Seldman, N.N.,(1984). Recycling from municipal refuse: A state of the Art Review and Annotated Bibliography, World Bank Technical paper N0 30. The World Bank, Washington, DC.

Environmental Law Research Institute (2009). Compilation of institutions and waste management regulations in Nigeria. Available from

http://www.elri-ng.org/newsandrelease2_waste.html

Fortelny, L., Michalkova,D & Kruli, Z (2004). An efficient method of material recycling of municipal plastic waste. *Polymer Degradation and stability,* 85 (3): 975 -979.

Garg, A., Kumar, K & Verma, V (2007). Public private partnership for solid waste management in Delhi: A case study. *Proceedings of the International Conference on sustainable solid waste management,* 5-7 September, Chennai, India, pp552 -559.

Huhtala A. A (1997). Post-consumer waste management model for determining optimal levels of recycling and landfilling. *Environ Resour Econ,* 10:301–314

Hoornweg, D., Thomas, L., & Otten, L., (1999). Composting and its applicability in developing countries. *Urban waste management working paper series 8.* Washington, DC; World Bank.

Imam, I., Mohammed, I.B., Wilson, D.C & Cheeseman, C.R (2008). Solid waste management in Abuja, Nigeria. *Waste Management* 28(2): 468-472.

Jang, J., Yoo, T., Oh, J & Iwasaki, I (1998). Discarded tire recycling practices in the United States, Japan and Korea. *Resources, Conservation and Recycling,* 22 (1-2): 1-14.

John, N. M., Adeoye, G. O. & Sridhar, M. K. C. (1996). Pelletization of compost: Nigerian Experience, *Biocycle*, USA, 7, pp. 53 -54.

Lavee, D & Khatib, M (2010). Benchmarking in municipal solid waste recycling. *Waste Management*, 2204 – 2208.

McCosh, K & Getliff, J. (2004). Effect of drilling fluid components on composting and the consequences for mud formulation. AADE Drilling Fluid Conference, Radisson Astrodome, Houston, Texas, AADE-04-DF-HO-25. April 6 -7. .Available from http://www.aade.org/TechPapers/2004Papers/Environmental%20Assurance/A ADE-04-DF-HO-25.pdf

Melendi, S., Diez, M.A., Alvarez, R and Barriocanal, C (2011). Plastic wastes, lube oils and carbochemical products as secondary feedstocks for blast-furnace coke production. Fuel Processing Technology, 92(3): 471-478.

Merrild, H., Damgaard, A., Christensen, T (2008). Life cycle assessment of waste paper management: The importance of technology data and system boundaries in assessing recycling & incineration. Resources, Conservation and recycling, 52: 1391-1398.

Occupational Health and Safety Administration (2008). Guidance for the identification and control of safety and health hazards in metal scrap recycling. Available from http://www.osha.gov/Publications/OSHA3348-metal-scrap-recycling.pdf Accessed 2nd August, 2011.

Ogwueleka, T.C (2009). Municipal solid waste characteristics and management in Nigeria. Iran. J. Environ. Health. Sci. Eng., 6 (3): 173-180.

Otti, V.I. (2011). J A model for solid waste management in Anambra State, Nigeria. Journal of Soil Science and Environmental Management, 2(2): 39-42.

Sampson, E & Etomi, G (2011). Environmental Legislation changes in Nigeria: what impact on foreign investment? Available from http://www.geplaw.com/media/Publications/Environmental%20Legislation%20 Changes%20in%20Nigeria.pdf?phpMyAdmin=17c4d67d23et18bb5773r72bc Accessed 1st August, 2011.

Selke, S.E (2002). Recycling, Handbook of plastics, Elastomers and composites, (4th edn), McGraw-Hill, New York, pp 693-757.

Wikipedia, the free encyclopedia. Available from http://en.wikipedia.org/wiki/Port_Harcourt Accessed 2nd August, 2011.

Wolsky, A.M & Gaines, L.L (1981). Discarded tires: A potential source of hydrocarbons to displace petroleum. Resource and Energy,3 (2): 195 – 206.

Yang, G.C.C (1993). Recycling of discarded tires in Taiwan. Resources, Conservation and Recycling, 9 (3):191-199

Zurbrugg, C., (2003). Solid waste management in developing countries. Available from http://www.eawag.ch/forschung/sandec/publikationen/swm/dl/basics_of_SW M.pdf Accessed 2nd August, 2011.

Zerbock, O (2003). Urban Solid Waste Management: Waste Reduction in Developing Nations. Available from http://www.cee.mtu.edu/sustainable_engineering/resources/technical/Waste_re duction_and_incineration_FINAL.pdf Accessed 2nd August, 2011.

9

Ballast Water and Sterilization of the Sea Water

María del Carmen Mingorance Rodríguez
Universidad de La Laguna
Spain

1. Introduction

Seventy one per cent of the earth's surface is covered by water; and ninety five point five per cent of the whole existing water is concentrated in oceans.

Generally, when we talk about water pollution or wastes dumped on sea, we think about products with varying degrees of toxicity (hydrocarbon, pesticides, etc.), or about innocuous products which, when their physicochemical conditions change and they are brought in a particular ecosystem, can be damaging to it (e.g.: a mass of water to a greater o lesser degree of salinity and/or temperature than the ones in the environment). However, the pollution and the alteration of an ecosystem can also be produced by the introduction of new allochthonous or foreign species.

Since human being started to sail, aquatic organisms have had the chance to travel in the hulls of vessel or in other surfaces: ropes, anchors, etc. (fouling). Therefore, those "unwanted travellers" have increased in last decades, with the recreational boat increase and with the transport of merchandise in great cargo ships, both in the hulls of vessel and into the ballast water.

If a vessel sails without cargo or partially loaded down, it will need ballast to keep its stability and safety.

Water is often used as ballast, but, when it is collected in origin, a series of organisms can travel into the water, normally members of the plankton community which could be stranger in the destination (allochthonous), and they might cause unwanted biogological pollution.

In order to avoid this possible pollution, we have to use a effective system to sterilize the ballast water.

Although the main methods to avoid the problems caused by ballast water have historically been mechanical, physical and chemical ones, on July 2007 the Marine Environment Protection Committee (MEPC) granted basic and final approval to a water sterilization system which does not use chemical products, based on the Advanced Oxidation Technology (AOT).

It's necessary the commitment of all the countries, the shipping companies, the industry and the scientific community to continue studying, developing and experimenting with efficient methods to sterilize the ballast water.

2. The planktonic community

Living beings that inhabit in oceanic waters can basically be classified in three communities: plankton, nekton and benthos. Normally, members of nekton and benthos are macroscopic, but members of plankton community are usually small and, by definition, they have little or no capacity to counteract the sea dynamic. That is why, until well into the 21st century, imagining an organism belonging to this community which was able to move enough in order to settle down in a different environment was difficult.

There is plankton in all the water of the earth. In seas and oceans, these animal and plant organisms, normally with a small size, are in every latitude and in the entire water column.

The components of this community have been clasified taking into acount several aspects or criteria, but not all the established classifications have a real rigor or scientific basis. Although, traditionally planktonic organisms have been divided into phytoplankton (autotrophs) and zooplankton (heterotrophic), in the latest classification, this distinction does not seem appropriate, since, among the autotrophic organisms, plants, some protists and bacteria are included, and there are animals, some protists and bacteria in the heterotrophic (Cognetti et al., 2001).

We can do a first classification in accordance with the permanence of the organisms in the community. As well, you can distinguish between holoplankton (Photo 1), formed by those organisms that throughout their life are part of the plankton, and meroplankton (Photo 2), constituted by those organisms that were part of the plankton during a period of their life and then became part of other communities (nekton or benthos).

In relation to its size, this type of classification was originally performed taking into account the dimensions of the mesh used for catches. Over time various authors have carried out these ratings, without reaching the final agreement, although currently, one of the most used is the classification proposed by Sieburth et al., 1978 (Table 1).

Category	Size	Organisms
Femtoplankton	0.02 -0,2 μm	Marine viruses
Picoplankton	0.2 – 2 μm	Bacteria; small eukaryotic protists
Nanoplankton	2 – 20 μm	Small flafellates; diatoms
Microplankton	20 - 200 μm	Foraminifera; rotifers; copepods nauplii
Mesoplankton	0.2 – 20 mm	Copepods; cladocera; ostracoda; chaetognaths; pteropods; tunicata; medusae
Macroplankton	2 – 20 cm	Heteropoda; chaetognaths; euphausiacea; medusae; salps; doliolids
Megaloplankton	20 – 200 cm	Ctenophores; salps; pyrosomes

Table 1. Classification of Planktonic organisms according to their size

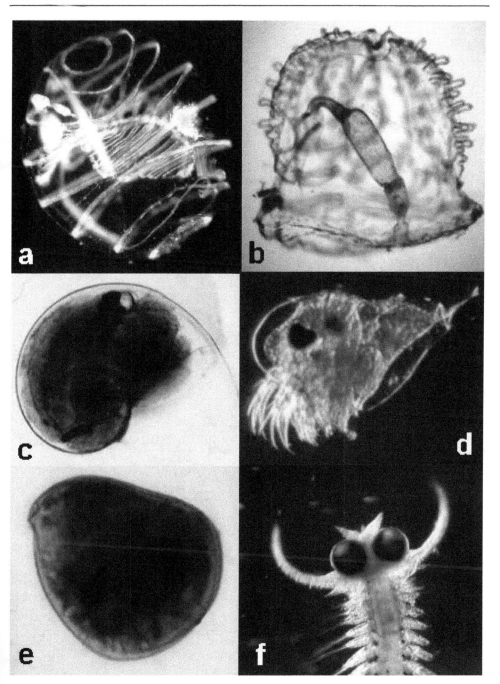

Photo 1. Components of the holoplankton: a) doliolid, b) hydromedusa; c) pteropod; d) cladocer; e) ostracod: and f) polychaeteo

The quantity and the variety of beings which are part of the planktonic community depend on a lot of factors: the latitude, the season, the bathymetric level, etc. In oligotrophic oceanic water, the vertical tows of 50 metre of length samplings were carried out until surface show average values of approximately 320 ind.m^{-3} (Mingorance et al., 2004).

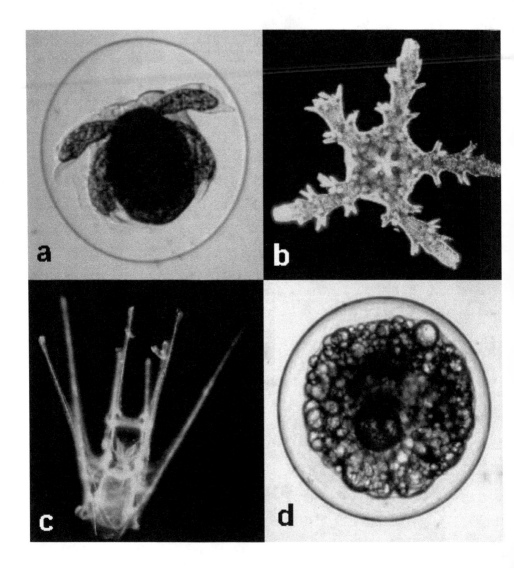

Photo 2. Components of the meroplankton: a) crustacea egg; b) ophiuroidea larvae; c) echinoderm pluteus larva; and d) fish egg

3. Biological pollution by ballast water

The first large enough boats to carry goods for trade appeared around 3500 B.C.; probably, the human being used boats of small size previously. From that moment the aquatic organisms could also travel: in the hulls of vessel (fouling), in the ropes, in the anchors, and so on (Photo 3).

Photo 3. The aquatic organisms have always travels in the hulls of vessel (fouling)

Nowadays, we try to minimize those "unwanted travellers" with the application of specific paints or with other methods like, for example, the utilization of ultrasounds. These measures are taken not only because they benefit environment but also because if the hull is clean, the vessel will have a greater energy saving and it will pick up more cruising speed.

The risk of possible biological pollutions could be caused not only by the organisms stuck in the hull of the vessel, but also because they travel accidentally in the water used to ballast the vessel when it goes without cargo.

The International Maritime Organization (IMO) define ballast water as "means water with its suspended matter taken on board a ship to control trim, list, draught, stability or stresses of the ship.".

The first materials used as ballast were stones, sand and other heavy objects, until sea water began to be used in the 19th century.

Throughout the 20th century and in this first decade of 21st century the transport by sea, both freight and passengers, has increased and the ships are becoming faster and with greater capacity. Moreover, the involuntary transportation of microorganisms has also increased among different places of the planet.

When cargo ships (bulk carriers, ore carries, tankers, among others) (Photo 4) must sail without cargo (with ballast) they need big volumes of ballast water, which is pumped before undertaking the trip with the empty vessel. If the ballast tanks are segregated, in other words, they transport exclusively sea water, and the vessel travel with ballast, organisms (virus, bacteria, animals and plants), normally belonging to the planktonic community, can accidentally go on board.

Photo 4. Carrier need ballast (water) when they travel without cargo in order to keep their stability

It's estimated that 10.000 million tons of ballast water are transfered among different places on the earth every year. Each vessel can transport from several hundred of liters until 100.000 tons of ballast water, it depends on its dimensions and purposes, and various studies suggest that the shipping industry is responsible for transfering more than 10 millions tons of ballast water every year among different places on the planet (Rigby et al., 1999).

Around 3000 species of animals and plants are estimated to travel daily by this way (IMO, 1998), and, though the survival rate of these species is low in a different habitats from the own one, as less than 3% of them really settle in new habitat, invading species are the direct cause of the 39% of known extinctions and of the loss of biodiversity.

As well, the speed of current vessels displacements is another factor that favors the prosperity of these organisms, because a lot of them wouldn't survive a long period of time in the darkness of the ballast tanks. That's why it's important that there aren't eggs, spores or organisms which can pollute that ecosystem in the ballast water throwed away in the destination place.

Ballast water is recognized as the most important vector of transoceanic and interoceanic movements of low deep water and of coastal organisms (IUCN, 2000).

This problem appeared for the first time before the IMO on 1988. In September of that year, Canada presented a report about a research related to the existence and the effects of foreign organisms (the Zebra Mussel, *Dreissena polymorpha*) in the vessel ballast unloaded in the Great Lakes between Canada and the United States from the Black Sea in the mid-1980s in the Marine Environment Protection Committee (MEPC), and, for more than 15 years, IMO has discussed about an agreement to regulate a better treatment of ballast water.

Throughout history, there were a lot of situations when these unexpected, unwanted and involuntary "travelers" have been detected.

In the port with the great traffic in the world, San Francisco Bay, 212 allochthonous species have been found (Garcia Varas, 2007).

In a bay in Oregon State (United States of America), 367 marine organisms from Japanese fauna were detected four years later than some ships from that country had released ballast water in the above-mentioned bay (UNESCO).

A planktonic Asian alga, specifically a diatom, which is able to reproduce very easily, invaded the North Sea on 1903, and it has been spreading to the Southern coast of United Kingdom (Boalch & Harbour, 1977).

The *Pfiesteria piscicida*, a dinoflagellate species of the genus *Pfiesteria*, which can have 24 different shapes and some of them produce a series of toxins, that are inocuous for human being but they are associated with injuries and with mortality of many fishes, was discovered on 1988 in North Carolina, after it had been brough into the ballast water (Rublee et al., 2005).

And, there are not only cases of planktonic beings: European green crab (*Carcinus maenas*) is a crustacean of about 8 centimeters long (in adult phase), it is quite voracious and practically omnivore. It has been introduced by ballast waters in Hawaii, both United States coasts, Panama, Madagascar, the Red Sea, India, Australia and Tasmania (Jaquenod de Zsögön, 2005).

In the United Nations Conference on Environment and Development (UNCED) held in Rio de Janeiro in 1992, an entire chapter of Agenda 21 was dedicated to the seas and oceans protection. Subsequently, the IMO announced, from 1993 to 1997, directives to avoid the transfer of unwanted aquatic organisms and pathogenic agents from ballast water and from sediment unloading, but, the problem is that, unlike hydrocarbon spillage and other forms of sea pollution caused by shipping traffic, marine species and exotic organisms, they are very difficult to eliminate.

The A.868(20) resolution, which was approved on November 27, 1997 as an appendix of the MARPOL agreement, entitled "Directives to the control and management of ballast water of vessels in order to minimize the transfer of damaging aquatic organisms and pathogenic agents", and its objective is to achieve that all the world fleet would sterilize the ballast water as soon as possible. In addition, it asks governments for promoting urgent measures to apply these directives and to spread them in shipping sector.

Members of the femtoplankton, picoplankton, nanoplankton, microplankton (Photo 5) and mesoplankton (Photo 6) (0.02 μm - 20 mm holoplankton), as well as of the meroplankton corresponding to the sizes, travel daily in the ballast water. That's why an efficient sterilization is so important.

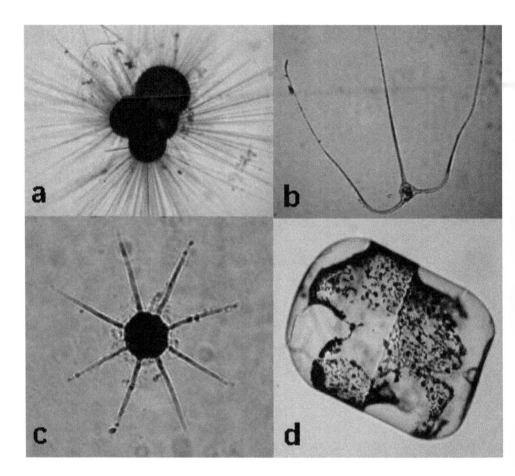

Photo 5. Microplankton: a) Globigerinida (foraminiferan); b) dinoflagellate; c) radiolarian; and d) diatom

According to a research done in North Tenerife (Canary Islands), a place is considered a oligotrophic area compared to the thickness of mesozooplankton which is in other places and other latitudes, an average thickness of 313.06 ind.m^{-3}.

A 60.000 tons oiler carries around 25.000 tons of ballast water in a travel with heavy sea, as 1 m^3 of sea water weights 1.020 g, 25.000 tons represent approximately 24.509.804 m^3. Therefore, if an oiler sucked up 24.000.000 m^3 of ballast water, it could pick up about 7.512 millions of organisms belonging to this community (Mingorance et al., 2009).

Photo 6. Copepods are usually the dominant members of the mesozooplankton

4. Sterilization of ballast water

Throughout time, mechanical methods as well as physical, chemical and biological ones have been considered, in order to solve the potentials problems of biological pollution, which could be caused by ballast water.

4.1 Mechanical methods

Mechanical methods can be the filtration, the reballasting, the dilution and the cyclonic separation, among others.

The filtration, which in the ballast water case need to be carried out as water is loaded on board, makes it possible to eliminate big particles as macroscopic algae, but it doesn't prevent from loading small organisms. Wastes would be left in the ballast intake area, but the necessary cost for this infrastructure could be very high.

Reballasting consist of doing the change of ballast water in deep waters, with 2000 meters or more sounding line, but it avoids the superficial water, the dredging areas and the places where there are illness or plankton outbreaks. This method keeps the chances of carrying allochthonous organisms at the place of destination.

Dilution caused by overflowing has to be done on the high sea; through the entry of water with the ballast pump and then we allow it overflows by the deck, at least a third of tank total volume. This method entails the risk of organisms wouldn't be entirely expelled.

In the cyclonic separation, ballast water goes into the inside of a chamber with circular flow; it goes through a venturi passage which is between the interior chamber and the separation chamber. The centrifugal helicoidal action propels particles toward the walls and it moves them to the chamber of sediments, they go through the clean water to outflow pipe. Sediments are constantly purged through the sediments return pipe.

4.2 Physical methods

The physical methods include the water treatment with ultraviolet radiation, heat treatment, ultrasonic treatment or water treatment with ions generated electronically.

The effect of the water treatment with ultraviolet radiation changes, it depends on the kind of organisms, because some of them are very resistant to UV radiation; it could be a very effective method if it were combined with the filtration. It has not toxic or damaging side effects for pipes, pumps or coating.

The heat treatment involves rising the ballast water temperature above 40° C for 8 minutes, since these conditions are lethal to practically all the organisms. Reaching this temperature depends on there being heating sources on board in order to treat the ballast water during the crossing.

The ultrasonic treatment for liquids uses high frequency energy to cause a vibration in the liquid. When the liquid is exposed to these vibrations, it causes cavitations (formation, expansion and implosion of microscopic gas bubbles in the liquid). As the ultrasonic energy goes into the liquid, the gas bubbles grown until they have a critical size and they implode. If the cavitation is sufficiently intense, it will break the cytoplasmic membranes.

The electrolytic generation of metallic ions, specifically copper and silver ions, has, in principle, a highly effectiveness to sterilize the water, but some organisms can increase their tolerance to high concentrations of copper and silver, so this turns the utilization of this method into a useless issue. Furthermore, the concentrations of these elements in the water could cause adverse environmental consequences. The application of this system has yet been rejected some time ago.

4.3 Chemical methods

Chemical treatments (basically the addition of disinfectants and biocides), the effect of which is oxidizing the organic matter, like, for example, chlorine, are refused for use, because water treated by this way keep certain biocide characteristics that could affect later other species.

4.4 Biological methods

On February 13, 2004, IMO adopted at London the Ballast Water Management Convention (BWM-2004) with the participation of 74 countries; the standards approved are compulsory from 2009 to 2016, depending on the year of vessel construction and the capacity of its ballast tanks. Therefore, vessel should have a specific ballast water management plan, which would be included in the documentation related to its operations.

Nevertheless, the 18th article of this Convention establish that it will come into effect 12 months later than the date when, at least, 30 states, whose merchant fleets represent the 35% or more of the world merchant fleet gross tonnage, will have ratified it.

Until February, 2010, 22 countries had ratified the Convention, but their fleets represented the 22.65% of the world merchant fleet gross tonnage.

In addition, this Convention won't be applied in the case of vessels which wasn't build for carrying ballast water, vessels that just carry ballast water in sealed tanks, vessels in a geographical area which only operate under the jurisdiction of that area, vessels which, although they were from another country, they operate under the jurisdiction of a geographical area or war vessels.

Since the approval of BWM-2004 Convention, the industry has been looking for solutions. It has mainly tried biological methods, which are effective and accessible from the point of view of economic costs, but also for their material installation on board. As well, according to these guidelines, the Swedish *Seatrade* awards to Countering Marine & Atmospheric Pollution have recognized two ballast water treatment systems: *PureBallast* on 2005 and *OceanSaver* on 2006.

In the *OceanSaver* System, during the ballast, the water goes through a 50 µm filter to eliminate the particles and organisms with a big size; this also helps to reduce the number of sediments that are piled up on the ballast tanks. Subsequently, the water is sterilized with a cavitation device, which breaks most of the organic matter, and then we add purify nitrogen and hydroxide ions generated by electrolysis.

In the *PureBallast* System, during the ballast, the water also goes through a 50 µm filter to eliminate the particles and organisms with a big size. Then, the water goes to the advanced oxidation unit (AOT unit), which contains catalyst of titanium dioxide and which, when is exposed to the light, it generate active oxygen and hydroxide radicals; the creation of those radicals doesn't depend on the water physico-chemical conditions (salinity, cloudiness, etc.); the radicals, whose life just last some milliseconds, break the cytoplasmic membrane of the organisms cells that have gone through the filter without using chemical products or generating damaging waste. The rest of the filtration is given back to sea, in the same place where ballast water was taken.

During the unballast, the water goes again through the advance oxidation unit to be sterilized again, but it doesn't go through the filter in order to there won't be possibility of unloading any kind of organic remains.

This treatment removes more than the 99.999% of the organisms with a size of more than 10 microns, and it can get the same percentage in the concentration of *Escherichia coli* (Alfa Laval, 2008). Thank to its modular design, it could adapt itself to several kinds of vessels and it can sterilize between 25 and 5.000 m³/h (MEPC, 2006).

The Marine Environment Protection Committee (MEPC), in the meeting celebrated from 9 to 13 July, 2007, at the Horticultural Halls, London, agreed on giving the basic and final approval *PureBallast* System, created by Alfa Laval, and proposed by Sweden and Norway.

5. Conclusion

More than 3000 species of bacteria, animals and plants travel daily into the ballast water, 10 million tons of ballast water are transported every year among different places on the planet and more than 7.500 millions of organisms can travel into the ballast water in a oiler carries of average size (60.000 tons). So it's indispensable to apply always the principle of prevention, and to achieve that water would be sterilize effectively, in order to avoid species germane to an ecosystem arrive and establish themself in another ecosystem, which would be different to the own one.

Mechanical, physical, chemical and biological methods have been developed and tested to minimize the transport of organisms into the ballast water, but, some of these methods have also turned out to be damaging to marine environment, because, when a problem was being solved, other ones appeared: for example, the introduction of chemical products in the environment.

The organisms transfer in the ballast water is, in the first decade of 21st century, the fourth most serious environmental problem worldwide. To avoid it, IMO approved a ballast water sterilization system based on the Advanced Oxidation Technology (AOT), which complies with the rules of this institution and which came into effect on 2009.

This agreement is very important for trying to avoid the arrival of invading species in the oceanic ecosystems; nevertheless, it's also necessary the commitment of all the countries, the shipping companies, the industry and the scientific community to study, develop, bring into operation and experiment with efficient methods to sterilize these waters.

6. Acknowledgment

I would like to thank to my daughter, Andrea Mingorance Hernández, for having translated into the English the original text from this chapter.

7. References

ALFA LAVAL. (2007). *PureBallast*, protección contra especies marinas indeseadas. *Rev. Rotación*, 447, pp. 28-30, ISSN 0211-2892.

ALFA LAVAL. (2008). *PureBallast* Proven Effectiveness. 02-08-2011, Available from http://www.alfalaval.com/pureballast

Boalch, G.T. & Harbour, D.S. (1977). Unusual diatom off the coast of south-west England and its effect on fishing. *Nature* 269, pp. 687-688, ISSN 0028-0836.

Cognetti, G.; Sará, M. & Magazzú, G. (2001). *Biología marina*. Ariel Ciencia. 624 pp, ISBN 978-84-344-8031-5.

Fonseca de Souza, M.H.; Leppakoski, E. & Librando, G. (2008). *The international law on ballast water*. Martinus Nijhoff Publishers. 401 pp. ISBN 978-90-04-16652-3.

García Varas, J.L. (2007). Polizones en las aguas de lastre amenazan los ecosistemas costeros. *Rev. WWF Panda*, pp. 18-19.

Horwood, J. W. & Driver, R. M. (1976). A note on a theoretical subsampling distribution of macroplankton. *J. Cons. Int. Explor. Mer*, 36 (3), pp. 274-276. ISSN 1054-3139.

IMO-MEPC. (1997). Results of ballast water exchange tests using the dilution method. *IMO MEPC 40th session*. London.

IMO/FAO/UNESCO-IOC/WMO/WHO/IAEA/UN/UNEP. Joint Group of Expert on the Scientific Aspects of Marine Environmental Protection (GESAMP). (1998). *Opportunistic settlers and the problem of the ctenophore Mnemiopsis leidyi invasion in the Black Sea*. IMO/UNEP. 84 pp. ISBN 92-801-1436-4. London (United Kingdom).

IMO. (2004). International Convention for the Control and Management of Ships' Ballast Water and Sediments. 02-08-2011. Available from http://www.imo.org/About/Conventions/ListOfConventions/Pages/Internation al-Convention-for-the-Control-and-Management-of-Ships'-Ballast-Water-and-Sediments-(BWM).aspx

IMO. (2005). *Ballast Water Management Convention*. International Maritime Organization. 141 pp. ISBN 92-801-0033-5.

IUCN. (2000). IUCN Guidelines for the Prevention of Biodiversity Loss Caused by Alien Invasive Species. *Information Paper of Fifth Meeting of the Conference of the Parties to the Convention on Biological Diversity*. Nairobi, Kenya.

Jaquenod de Zsögön, S. (2005). *Derecho Ambiental. La Gobernanza de las Aguas*. Ed. Dykinson, S.L. 266 pp. ISBN 9788497727396.

MEPC. (2006). Brief Description of the Technologies Presented. *Report of the Marine Environment Protection Committee, October 2006*. MEPC 55/WP.4 Annex 1.

Mingorance, Mª. C.; Lozano Soldevilla, F.; Braun, J. A.; Landeira, J. Mª; Espinosa, J. Mª. & Gómez, J. I.(2004). Estudio de la distribución vertical de la comunidad mesozooplanctónica en aguas de la isla de Tenerife (Islas Canarias). *Rev. Acad. Can. Cienc.*, XV (3-4), pp. 99-114. ISSN 1130-4723.

Mingorance, Mª. C.; Gómez, J. I.; Lozano, F.; Gómez, A. U.; González, J. A. & Calvilla, J. M. (2009). Ballast and unballast operations in oil tankers: Planktonic organisms that can travel with the ballast water. *Journal of Maritime Research*, Vol. VI (3), pp. 27 -39. ISSN 1697-4840.

Primac, R. B. & Ros, J. (2002). *Introducción a la biología de la conservación*. Ariel Ciencia. 288 pp. ISBN: 978-84-344-8039-1.

Rigby, G.R.; Hallegraeff, G.M. & Sutton, C. (1999). Novel ballast water heating technique offers cost-effective treatment to reduce the risk of global transport of harmful marine organisms. *Mar. Ecol. Prog. Ser.*, Vol. 191, pp. 289-293. ISSN (printed) 0171-8630.

Rublee, P.A.; Remington, D.L.; Schaefer, E.F. & Marshall, M.M. (2005). Detection of the Dinozoans Pfiesteria piscicida and P. shumwayae: A review of detection methods and geographic distribution. *J. Euk. Microbiol*. 52 (2), pp. 83–89. ISSN 1066-5234.

Sieburt, J. M.; Smetacek, V. & Lenz, J.(1978). Pelagic ecosystem structure: Heterotrophic compartments of the plankton and their relationship to plankton size fractions. *Limnol. and Oceanog.*, 23, pp. 1256-1263. ISSN: 00243590.

Steele, J. H. & Frost, B. W. (1977). The structure of plankton communities. *Phil. Trans. R. Soc. London*, 280. pp. 485-534. ISSN: 1471-2970.

Williams, R.; Griffiths, F. B.; Van der Wal, E. J. & Kelly, J. (1988). Cargo vessel ballast water as a vector for the transport of non-indigenous marine species. *Estuarine, Coastal and Shelf Science*, Vol. 26 (4), pp. 409 – 420. ISSN: 0272-7714.

Part 4

Emissions Related to Waste Disposal

10

Evaluation of Replacing Natural Gas Heat Plant with a Biomass Heat Plant – A Technical Review of Greenhouse Gas Emission Trade-Offs

James G. Droppo and Xiao-Ying Yu
Pacific Northwest National Laboratory
USA

1. Introduction

Proposed fuel conversions can involve more than a simple reduction of emission rates. For example, a Renewable Fuel Heating Plant (RFHP) was proposed for the U.S. Department of Energy (DOE) National Renewable Energy Laboratory (NREL) to replace a natural gas plant. The proposed RFHP replacement plant was to use a biomass fuel, wood chips. A review was conducted to address questions related to how increases in the plant's carbon dioxide (CO_2) emission rates could represent a desirable outcome. This review and its results are published here in the hope that it may be useful to others considering similar conversions. This chapter addresses 1) why despite an increase in emission rate the conversion is considered an effective reduction in greenhouse gas emissions, and 2) how the proposed wood chip combustion process emissions compare with other means of disposing of or using wood chips.

The 2001 and 2007 Assessment Reports of the Intergovernmental Panel on Climate Change (IPCC) considers the evidence for global climate change and the potential consequences of such changes [IPCC, 2001; 2007]. Based on the result of worldwide research efforts, this paper concludes that the earth's climate has changed over the last century. It also notes that there is recent strong evidence that human activities have caused most of the warming observed in the last 50 years. The current computer models are predicting that this temperature rise should continue over this century.

In terms of the net CO_2 in the atmosphere, the argument is made based on current scientific understanding on climate change processes, that burning of wood chips is much more desirable than a fuel that contains carbon that has been sequestered underground. The CO_2 from wood chip combustion has a "net zero" emission rate based on factors in the Environmental Protection Agency's AP-42. The "net zero" emission rate is based on an assumption that CO_2 from burning wood from forests represents no increase in the net amount of CO_2. A cycling of carbon between the atmosphere and forests results in no net gain or loss of airborne CO_2. On the other hand CO_2 from burning natural gas represents an increase in the net amount of CO_2 from the introduction of "new" carbon that has been previously sequestered underground. Thus argument for the proposed conversion is to stop the introduction of the new carbon into the current atmospheric carbon cycle.

From the viewpoint of minimizing impacts on global climate change, the burning of wood chips also tends to be more desirable than the common alterative use of wood chips in composting activities. Although there is great variability and uncertainty in the published emission rates, the gaseous emission from both open burning and composting tend to have much larger emissions of greenhouse gases, and specifically larger fractions of gases such as methane (CH_4) and ammonia (NH_3) than the proposed process for burning the wood chips. The published source terms for open burning show the incineration option to be preferable from the viewpoint of having lower emissions. Of particular importance are mixtures of combustion products from these activities. For example because methane is currently thought to be many times more effective for inducing climate changes than CO_2, the potentially higher methane levels from open burning and composting make these activities less desirable from the viewpoint of minimizing the potential impact of greenhouse gas emissions.

The paradox in the proposed conversion is that from an absolute quantity perspective the RFHP would emit more CO_2 than what is being currently emitted with natural gas firing. Although the main thrust in reducing atmospheric levels of greenhouse gases has been to reduce introduction of "new" carbon by the combustion of fossil fuels, some efforts have considered the possibility of combustion control strategies for agricultural and forestry products.

A review was conducted of recent literature relevant to these issues. The predominance of current literature points to a need to reduce greenhouse emissions, and it is assumed for the purpose of this review to be a reasonable basis for proceeding with actions that will reduce those emissions. The results of this review are reported below. A form of these results were posted as a DOE report, we would like to make it avaiable in the public domain for people who are interested in this topic.

2. Background

The Third and Fourth Assessment Reports of the IPCC [2001, 2007] considers the evidence for global climate change and the potential consequences of such changes[1]. Based on the result of worldwide research efforts, the report concludes that the earth's climate has changed over the last century. These reports also notes that there is mounting evidence that human activities have caused most of the warming observed in the past 50 years. The current computer models predict that this temperature rise should continue over this century.

The IPCC reports note that changes in climate are the result of both internal variability within the climate system and external factors (both natural and anthropogenic). Human emissions are significantly modifying the concentrations of some gases in the atmosphere. Some of these gases are expected to affect the climate by changing the earth's radiative balance, measured in terms of radiative forcing.

The 2007 IPCC reports provide an overview of the global effects of greenhouse gases and conclude that they tend to warm the earth surface by absorbing some of the infrared radiation it emits.

---

[1] The reader is referred to the websites http://www.ipcc.ch/ and "http://www.greenfacts.org" for additional information. The latter site provides summaries as well as quotes from the IPCC (2007) report.

"The principal anthropogenic greenhouse gas is carbon dioxide (CO_2), whose concentration has increased by 31% since 1750 to a level which is likely to have not been exceeded for 20 million years. This increase is predominantly due to fossil fuel burning, but also to land-use change, especially deforestation. The other significant anthropogenic greenhouse gases are CH_4 (151% increase since 1750, 1/3 of CO_2's radiative forcing), halocarbons such as CFCs and their substitutes (100% anthropogenic, 1/4 of CO_2's radiative forcing) and nitrous oxide (N_2O) (17% increase since 1750, 1/10 of CO_2's radiative forcing)."

The IPCC [2007] attributes that about three-quarters of the anthropogenic emissions of CO_2 to the atmosphere during the past 20 years is due to fossil fuel burning. The rest is attributed to predominantly land-use change, especially deforestation.

"Currently the ocean and the land together are seen as taking up about half of the anthropogenic CO_2 emissions. On land, the uptake of anthropogenic CO_2 is thought to very likely exceed the release of CO_2 by deforestation during the 1990s. The rate of increase of atmospheric CO_2 concentration has been about 1.5 ppm (0.4%) per year over the past two decades. During the 1990s the year to year increase varied from 0.9 ppm (0.2%) to 2.8 ppm (0.8%). A large part of this variability is due to the effect of climate variability (e.g., El Niño events) on CO_2 uptake and release by land and oceans."

Studies of the signatures of emissions from biomass and fossil fuel burning conducted by Reiner et al. [2001] over the tropical Indian Ocean provides some insight into the relative source importance. In the air from the continent they found that most of the CO is from biomass/biofuels burning and the majority of the aerosols are from fossil fuel burning. These results underscore the apparent importance of incomplete combustion products in the current emissions from combustion of biomass/biofuels.

The literature indicates that there is a world-wide effort to define means and methods of using bio-energy while minimizing greenhouse gas emissions. Faaij [2006] states that bio-energy is one of the key options to mitigate greenhouse gas emissions and substitute fossil fuels. The efforts are reflected in the wide range of activities and programs for developing and stimulating bio-energy.

3. Emission rates

This review focuses on greenhouse gas emissions from a proposed wood fired boiler and their existing natural gas boiler and from other alternative uses of the wood chips. The current and proposed emissions rates as well as the assumed annual tonnage of wood consumed are provided by DOE/NREL. A literature search was conducted with the objective of quantifying annual air emissions that would result from the likely alternative uses of the wood proposed to be used by PRFHP. The three alternatives considered were composting, land filling and open burning. Because each of these alternatives represents a wide range of processes, it is expected that there will be correspondingly a wide range of potential emissions from each of these alternatives.

As a result of a worldwide concern over the potential effect of greenhouse gases, the current literature contains considerable information on the potential emissions. These papers include process-specific studies of the emissions from biomass burning and other biofuel combustion processes (i.e. Borgwardt, 1997; Turn et al., 1997; Dennis et al., 2002; Hays et al.[2], 2002; Kasische and Penner, 2004; Hayes et al., 2005; Pronobis, 2006; Hashaikeh et al., 2007; Tilman et al., 2007] as well as quantification of global emissions from conventional and biogenic processes (i.e. Oros and Simonett, 2001; Ito and Penner, 2004; Liebig, 2005; Wiedinmyer et al., 2006[3]; Schmid et al., 2006[4]; Khalil et al.,2007[5]). The attached list of citations for papers considered in this review contains additional references concerning both process and global emission rates.

The emphasis of this review is to compare the emissions from natural gas and the wood fired boiler with comparable emission numbers for the same annual volume of wood disposed of by the alternative means. Specifically the comparison is based on the wood is either 1) processed into compost; 2) dumped in a landfill; or 3) subjected to open burning à la the forest service's preferred method for slash pile disposal.

Winiwarter [2001] detailed evaluation shows that "much of the overall uncertainty derives from a lack of understanding of the processes associated with N_2O emissions from soils. Other important contributors to greenhouse gas emission uncertainties are CH_4 from landfills and forests as CO_2 sinks. The uncertainty of the trend has been determined at near 5% points, with solid waste production (landfills) having the strongest contribution."

3.1 Proposed heat plant emissions

Table 1 shows a comparison of the emissions from the current natural gas and proposed biomass heat plants. Annual emissions from the proposed heat plant are based on a permitted 3,800 tons of biomass, at approximately 6,500 Btu/lb heating value. This maximum annual fuel consumption is based on the assumption that biomass has thirty to forty percent moisture. Uncontrolled emissions are based on emission factors referenced in the United States Environmental Protection Agency's (U.S. EPA's) AP-42 Compilation of Air Pollutant Emission Factors (5th edition), Chapter 1.6 *Wood Residue Combustion in Boilers*.

Combustion Scenario	Current Heat Plant	Woodchip Heat Plant
Air Pollutant	Existing Emissions using Natural Gas Boilders	Wood Combustion
Units	tons/yr	tons/yr
Carbon Monoxide (CO)	1.69	3.58
Sulfur Dioxide (SO_2)	0.012	0.5
Nitrogen Oxide (NO_x)	3.3	4.3
PM Total	0.1	3.2
VOC (non methane)	0.1	0.1
Carbon Dioxide (CO_2)	2340	Net zero

Table 1. Emissions Comparison of Current Natural Gas and Proposed Biomass Heat Plants

---

[2] Considers emissions from burning of several types of wood.
[3] Estimation of emissions from fires in North America.
[4] Considers carbon budget for forests.
[5] Recent summary of importance of atmospheric methane as a greenhouse gas.

3.1.1 Composting

For all compositing activities, the products produced depend heavily on conditions. High (i.e. greater than 50%) moisture content is a prerequisite for having high levels of microbial activities [Lang et al., 2006]. Dispite the wide range of possible emissions; it is possible to describe general trends from composting.

The aerobic composting of chips from clear-cut trees is considered by Suzuk et al. [2004]. They found that certain combinations of materials could be composed within 10 months time period. Although they did not specifically consider the emissions, it is likely that emissions of CO_2 and CH_4 are much lower than what would have occurred with anaerobic composting.

IPCC (2007) states that:

"Composting refers to the aerobic digestion of organic waste. The decomposed residue, if free from contaminants, can be used as a soil conditioner. As noted above under landfilling, greenhouse gas emissions from composting are comparable to landfilling for yard waste, and lower than landfilling for food waste. These estimates do not include the benefits of the reduced need for synthetic fertilizer, which is associated with large CO_2 emissions during manufacture and transport, and N_2O releases during use. USDA research indicates that compost usage can reduce fertilizer requirements by at least 20% [Ligon, 1999], thereby significantly reducing net greenhouse gas emissions.

Composting of yard waste has become widespread in many developed countries, and some communities compost food waste as well. Small, low-technology facilities handling only yard waste are inexpensive and generally problem-free. Some European and North American cities have encountered difficulties implementing large-scale, mixed domestic, commercial and industrial bio-waste collection and composting schemes. The problems range from odor complaints to heavy metal contamination of the decomposed residue. Also, large-scale composting requires mechanical aeration which can be energy intensive (40-70 kW/t of waste) [Faaij et al., 1998]. However, facilities that combine anaerobic and aerobic digestion are able to provide this energy from self-supplied methane. If 25% or more of the waste is digested anaerobically the system can be self-sufficient [Edelmann and Schleiss, 1999].

For developing countries, the low cost and simplicity of composting, and the high organic content of the waste stream make small-scale composting a promising solution. Increased composting of municipal waste can reduce waste management costs and emissions, while creating employment and other public health benefits.

Anaerobic digestion to produce methane for fuel has been successful on a variety of scales in developed and developing countries. The rural biogas programmes based upon manure and agricultural waste in India and China are very extensive. In industrial countries, digestion at large facilities utilizes raw materials including organic waste from agriculture, sewage sludge, kitchens, slaughterhouses, and food processing industries. "

3.1.2 Landfill

Disposal of wood chips in a landfill can result in a wide range of greenhouse gas emissions. Although conditions in a land fill are typically anaerobic, it is also possible to have aerobic conditions. The majority of current research articles report studies based on controlling mixtures and conditions to minimize greenhouse emissions.

Pier and Kelly [1997] indicate that as a result of it high degradable organic carbon content of wood wastes combined with a tendency for anaerobic onsite storage conditions, wood wastes are a potent potential source of methane production. They also indicate that the emissions from such wastes would be much higher if such stored wastes were put in soil-capped landfills. CO_2 and methane emission rates were measured in terms of mass emitted per surface area of a sawdust pile. They found methane flux rates that were within, but at the low end of the range of landfill methane emission values, as reported in the literature.

Traditional landfills tend to be significant sources of greenhouse gases. If one assumes similar processes will occur for the wood as for the other organic materials in the landfill, then the disposal of wood chips in a landfill will result in significant releases of potent greenhouse gases – and thus from the viewpoint of the potential to generate greenhouse gas, burial in a traditional landfill will be a less desirable fate for the wood chips than the proposed incineration process.

Typical landfill gas composition includes 63.8% CH_4, 33.6% CO_2, 0.16% O_2, 2.4% nitrogen (N_2), 0.05 % hydrogen (H_2), 0.001% carbon monoxide (CO), 0.005% ethane (C_2H_6), 0.018% ethane (C_2H_4), 0.005% acetaldehyde (C_2H_5O), 0.002% propane (C_3H_8), 0.003% butanes (C_3H_8), 0.00005% helium (He), < 0.05% higher alkanes, 0.009% unsaturated hydrocarbons, 0.00002% halogenated compounds, 0.00002% hydrogen sulphide (H_2S), 0.00001% organosulphur compounds, 0.00001% alcohols, and 0.00005% other compounds [Al-Dabbas, 1998]. Although recent work by Launghi et al. and references therein suggested different percentages of the landfill gas composition in Europe, CH_4 content is highest among landfill gas emission [Lunghi et al., 2004]. Based on the Italian report, CH_4 accounts for 58.01%, CO_2, 41.38%, O_2, 0.13%, N_2 0.48%, and H_2O 0.41%.

Methane can be formed by various paths including biogenic (bacterial) methane formation, thermogenic formation, and incomplete combustion of biomass or fossil fuels. After formation of CH_4, it can be modified by secondary processes. One of the most important processes is aerobic CH_4 oxidation by methanotrophic bacteria, which occurs in many natural and anthropogenic environments producing CH_4 under anaerobic conditions. Substantial aerobic methane oxidation has been found in various environments such as natural wetlands, rice paddies and landfill sites [Bergamaschi et al., 1998].

Emission of CH_4 from landfills is causing increasing concerns in global climate change, because its warming potential is 20 times higher than that of CO_2 in a time trajectory of 100 years [Kumar et al., 2004a; Kumer et al., 2004b]. Formation of organic compounds including degradable materials with high molecular weights is closely linked with emission of CH_4 in landfills [Pan and Voulvoulis, 2007].

IPCC (2007) states that:

"Worldwide, the dominant methods of waste disposal are landfills and open dumps. Although these disposal methods often have lower first costs, they may contribute to serious local air and water pollution, and release high GWP landfill gas (LFG). LFG is generated when organic material decomposes anaerobically. It comprises approximately 50%-60% methane, 40%-45% CO_2 and the traces of non-methane volatile organics and halogenated organics. In 1995, US, landfill methane emissions of 64 MtC_{eq} slightly exceed its agricultural sector methane from livestock and manure.

Methane emission from landfills varies considerably depending on the waste characteristics (composition, density, particle size), moisture content, nutrients, microbes, temperature, and pH [El-Fadel, 1998]. Data from field studies conducted worldwide indicate that landfill methane production may range over six orders of magnitude (between 0.003-3000g/m^2/day) [Bogner et al., 1985]. Not all landfill methane is emitted into the air; some is stored in the landfill and part is oxidized to CO_2. The IPCC theoretical approach for methane estimation has been complemented with more recent, site-specific models that take into account local conditions such as soil type, climate, and methane oxidation rates to calculate overall methane emissions [Bogner et al., 1998].

Laboratory experiments suggest that a fraction of the carbon in landfilled organic waste may be sequestered indefinitely in landfills depending upon local conditions. *However, there are no plausible scenarios in which landfilling minimizes greenhouse gas emissions from waste management (italics added).* For yard waste, greenhouse gas emissions are roughly comparable from landfilling and composting; for food waste, composting yields significantly lower emissions than landfilling. For paper waste, landfilling causes higher greenhouse gas emissions than either recycling or incineration with energy recovery [US EPA, 2000]."

The potential for the gases produced in landfills to be collected as a biomass source is being considered [McKendry, 2002 a;b;c] as a process that potentially could put the landfill greenhouse emissions on a par with the proposed heat plant.

3.1.3 Open burning

The disposal of the wood chips by open burning is considered. Fires produce gases and aerosols to the atmosphere such as CO_2, CO, nitrogen oxides (NO_x), volatile and semivolatile organic compounds (VOC and SVOC), particulate matter (PM), NH_3, sulfur dioxide (SO_2), and CH_4 [Clinton et al., 2006].

The literature on biomass combustion encompasses biofuels (wood, crop waste, and dug-cake) and forest fires (accidental, shifting, cultivation, and controlled burning). In a study based in India, forest fires were reported as only being a 7% portion of the 93% of total combustion sources from biomass consumption [Reddy and Venkataraman, 2002]. Greenhouse gas emission from biofuel burning is much less in comparison to emission from open vegetation fires [Ito and Penner, 2005].

Open burning produces a much more complex mix of gas emissions. Hays et al. examined fine particulate matter ($PM_{2.5}$) and gas phase emissions from open burning of six fine foliar fuels commonly found in fire-prone ecosystems in the U.S. [Hays et al., 2002]. They identified more than 100 individual organic compounds in the fine carbonaceous particulate matter using gas chromatography/mass spectrometry. The emission ranges by organic compound class are the following, n-alkane 0.1-2%, ploycyclic aromatic hydrocarbon (PAH) 0.02-0.2%, n-alkanoic acid 1-3%) n-alkanedioic acid 0.06-0.3%, n-alkenoic acid 0.3-3%, resin acid 0.5-6%, triterpenoid 0.2-0.5%, methoxyphenol 0.5-3%, and phytoseterol 0.2–0.6%.

Table 2 compares the relative emissions from the proposed heat plant and the alternative of open burning of the wood chips based on emission factors listed in US EPA guidance [US EPA, 2008].

Combustion Scenario	Current Heat Plant	Proposed Heat Plant	Equivalent Mass of Wood
Air Pollutant	Existing Emissions Using Natural Gas Boilers	Wood Combustion	Open Burning
Units	tons/yr	tons/yr	tons/yr
Carbon Monoxide (CO)	1.69	3.58	270. (b) 170-370 (c)
Sulfur Dioxide (SO₂)	0.012	0.5	Not listed
Nitrogen Oxide (NOₓ)	3.3	4.3	7.7 (b)
PM Total	0.1	3.2	32. (b) 7.6 – 32.(c)
VOC (non methane)	0.1	0.1	45.8 (b) 7.6-36. (c)
Carbon Dioxide (CO₂)	2340	Net zero	Net Zero (Not Listed)
Methane (CH₄)	(a)	(a)	6.2-10.8 (c)

(a) no value listed, expected to be negligible
(b) estimated based on Rocky Mountain wildfire forest burning emission factors (AP-42, Table 13.1-3)
(c) estimated based on forest wastes burning emission factors (AP-42, Table 2.5-5)
(d) can be a significant source depending on conditions in the landfill.
(e) can be a significant source depending on the composting conditions.

Table 2. Comparison of Emissions: Current Plant, Proposed Plant, and Open Burning

Mouillot et al. [2006] concluded that "The total amount of carbon emitted to the atmosphere from biomass burning is uncertain." Neither combustion efficiencies nor the extent of burned areas are known with precision [Ito and Penner, 2005; Kasischke and Penner, 2004].

The argument for increased use of biofuels instead of fossil fuels is based on considering CO_2 from biomass burning as recyclable. An obvious question is whether the use of biofuels should be reduced as a strategy for reducing the total emissions of greenhouse gases. In fact, researchers predict that the effect of controls on biomass burning on climate change will be mildly effective in reducing CO_2 emissions – leading to a prediction of a short-term warming followed by a longer-term global cooling [Jacobson, 2004]. Although the Kyoto Protocol did not consider biomass-burning controls as a means of reducing global warming, there is an argument that at least in the time-frame of the next decade this strategy has the potential of reducing global warming.

3.2 Waste to energy facilities

The emission rates from waste to energy combustion in the proposed biomass heat plant are likely to be similar to those from incineration as opposed to those discussed above for open burning, landfilling, and composting.

ITCC [2007] states:

> "Incineration is common in the industrialized regions of Europe, Japan and the northeastern USA where space limitations, high land costs, and political opposition to locating landfills in communities limit land disposal. In developing countries, low land and labor costs, the lack of high heat value materials such as paper and plastic in the waste stream, and the high capital cost of incinerators have discouraged waste combustion as an option.
>
> Waste-to-energy (WTE) plants create heat and electricity from burning mixed solid waste. Because of high corrosion in the boilers, the steam temperature in WTE plants is less than 400 degrees Celsius. As a result, total system efficiency of WTE plants is only between 12%–24% [Faaj et al., 1998; US EPA, 1998; Swithenbank and Nasserzadeh, 1997].
>
> Net greenhouse gas emissions from WTE facilities are usually low and comparable to those from biomass energy systems, because electricity and heat are generated largely from photosynthetically produced paper, yard waste, and organic garbage rather than from fossil fuels. Only the combustion of fossil fuel based waste such as plastics and synthetic fabrics contribute to net greenhouse gas releases, but recycling of these materials generally produces even lower emissions."

Based on a broader life-cycle perspective, Borjesson and Berglund [Borjesson and Berglund, 2006; Berglund and Borjesson, 2006] address the overall environmental impact when biogas systems are introduced to replace various conventional systems for energy generation, waste management and agricultural production. Their conclusion is that biogas systems normally also lead to indirect environmental improvements, which in some cases are considerable. They note that these indirect benefits (e.g. reduced nitrogen leaching, emissions of ammonia and methane) often exceed the direct environmental benefits achieved in situation when fossil fuels are to be replaced by biogas.

4. Conclusions

Conversion of the NREL heat plant will replace combustion of fossil fuels with a biomass fuel. The literature was found to support this action in terms of emissions of greenhouse gases.

Although the reduction of any of the major emission sources can be an effective strategy for slowing the increase in atmospheric greenhouse gases, the literature also strongly supports the replacement of fossil fuels with biomass fuels based on the concept of stopping the release of "new" carbon currently sequestered in fossil fuels.

When compared to open burning, the proposed conversion is desirable in terms of the emission of pollutants. The comparison in Table 2 based on EPA's AP-42 indicates the same mass wood chips by open burning would significantly increase the emissions of carbon monoxide, total particulate matter, and VOCs. The literature indicates only a factor of 2 increase in nitrogen oxide emissions.

When compared to disposal in a landfill, the evidence is also in favor of the proposed conversion. Disposal landfills have the advantage over the biomass heat plant in that the landfill will generally not be a significant source of particulate matter. On the other hand, wood chips in a landfill with actively decomposing materials would produce a more potent mixture of greenhouse gasses (i.e. increased levels of methane and ammonia). The potential for the gases produced in landfills to be collected as a biomass source is considered as a process that potentially in the future could put the landfill greenhouse emissions on a par with the proposed heat plant [McKendry, 2002a]. IPCC [2007] indicates that there are no plausible scenarios in which landfilling minimizes greenhouse gas emissions from waste management.

When compared to composting, the proposed conversion is desirable in terms of the emission of pollutants. Although the literature is full of studies to make the emissions from composting more "greenhouse gas friendly," the current status is that a composting technology for organic material such as wood chips only minimizes the emissions of greenhouse gases. The high temperature combustion process used in the proposed biomass heat plant mainly produces CO_2 while nearly eliminating the emissions of the other more potent greenhouse gases.

Thus the concept of reducing the potential global warming impacts from burning of biomass is consistent with the proposed conversion of the heat plant. As seen in this review, in addition to using a recycled source of carbon, the high temperature combustion of the wood chips results in a net decrease in the climate-change potency of the greenhouse gas emissions compared to current use/destruction of wood chips in composting and open burning.

5. References

Al-Dabbas, M. A. F. (1998), Reduction of methane emissions and utilization of municipal waste for energy in Amman, *Renewable Energy*, 14(1-4), 427-434.

Bergamaschi, P., C. Lubina, R. Konigstedt, et al. (1998), Stable isotopic signatures (delta C-13, delta D) of methane from European landfill sites, *Journal of Geophysical Research-Atmospheres*, 103(D7), 8251-8265.

Berglund, M., and P. Borjesson (2006), Assessment of energy performance in the life-cycle of biogas production, *Biomass & Bioenergy*, 30(3), 254-266.

Bogner, J., K. Spokas, E. Burton, et al. (1995), LANDFILLS AS ATMOSPHERIC METHANE SOURCES AND SINKS, *Chemosphere*, 31(9), 4119-4130.

Bogner, J. E., K. A. Spokas, and E. A. Burton (1999), Temporal variations in greenhouse gas emissions at a midlatitude landfill, *Journal of Environmental Quality*, 28(1), 278-288.

Borgwardt, R. H. (1997), Biomass and natural gas as co-feedstocks for production of fuel for fuel-cell vehicles, *Biomass & Bioenergy*, 12(5), 333-345.

Borjesson, P., and M. Berglund (2006), Environmental systems analysis of biogas systems - Part 1: Fuel-cycle emissions, *Biomass & Bioenergy*, 30(5), 469-485.

Clinton, N. E., P. Gong, and K. Scott (2006), Quantification of pollutants emitted from very large wildland fires in Southern California, USA, *Atmospheric Environment*, 40(20), 3686-3695.

Dennis, A., M. Fraser, S. Anderson, et al. (2002), Air pollutant emissions associated with forest, grassland, and agricultural burning in Texas, *Atmospheric Environment*, 36(23), 3779-3792.

Faij, A. P. C. (2006). "Bio-energy in Europe: changing technology choices." *Energy Policy*, 34(3): 322-342.

Hays, M. D., C. D. Geron, K. J. Linna, et al. (2002), Speciation of gas-phase and fine particle emissions from burning of foliar fuels, *Environmental Science & Technology*, 36(11), 2281-2295.

Hays, M. D., P. M. Fine, C. D. Geron, et al. (2005), Open burning of agricultural biomass: Physical and chemical properties of particle-phase emissions, *Atmospheric Environment*, 39(36), 6747-6764.

IPCC, 2007. Climate Change 2007: The Physical Science Basis: Summary for Policymakers. Contribution of Working Group I to the Fourth Assessment Report of the Intergovernmental Panel on Climate Change. Available at http://www.ipcc.ch, last accessed May 4, 2007.

IPCC, 2001. Climate Change 2001: Mitigation. Contribution of Working Group III to the Third Assessment Report of the Intergovernmental Panel on Climate Change. Available at http://www.ipcc.ch, last accessed May 4, 2007.

Ito, A, and J. E. Penner (2005), Historical emissions of carbonaceous aerosols from biomass and fossil fuel burning for the period 1870-2000, *Global Biogeochemical Cycles*, 19(2).

Ito, A., and J. E. Penner (2004), Global estimates of biomass burning emissions based on satellite imagery for the year 2000, *Journal of Geophysical Research-Atmospheres*, 109(D14).

Jacobson, M. Z. (2004), The short-term cooling but long-term global warming due to biomass burning, *J. Clim.*, 17(15): 2909-2926.

Kasischke, E. S., and J. E. Penner (2004), Improving global estimates of atmospheric emissions from biomass burning, *Journal of Geophysical Research-Atmospheres*, 109(D14).

Khalil, M. A. K., and R. A. Rasmussen (2003), Tracers of wood smoke, *Atmospheric Environment*, 37(9-10), 1211-1222.

Kumar, S., S. A. Gaikwad, A. V. Shekdar, et al. (2004a), Estimation method for national methane emission from solid waste landfills, *Atmospheric Environment*, 38(21), 3481-3487.

Kumar, S., A. N. Mondal, S. A. Gaikwad, et al. (2004b), Qualitative assessment of methane emission inventory from municipal solid waste disposal sites: a case study, *Atmospheric Environment*, 38(29), 4921-4929.

Lang, D. J., C. R. Binder, R. W. Scholz, et al. (2006), Impact factors and regulatory mechanisms for material flow management: Integrating stakeholder and scientific perspectives - The case of bio-waste delivery, *Resources Conservation and Recycling*, 47(2), 101-132.

Liebig, M. A., J. A. Morgan, J. D. Reeder, et al. (2005), Greenhouse gas contributions and mitigation potential of agricultural practices in northwestern USA and western Canada, *Soil & Tillage Research*, 83(1), 25-52.

Ligon, J. M., D. S. Hill, P. Hammer, et al. (1999), Natural products with antimicrobial activity from Pseudomonas biocontrol bacteria, in *Pesticide Chemistry and Bioscience: The Food-Environment Challenge*, edited by G. T. Brooks and T. R. Roberts, pp. 179-189.

Lunghi, P., R. Bove, and U. Desideri (2004), Life-cycle-assessment of fuel-cells-based landfill-gas energy conversion technologies, *Journal of Power Sources*, 131(1-2), 120-126.

McKendry, P. (2002a), Energy production from biomass (part 1): overview of biomass, *Bioresource Technology*, 83(1): 37-46.

McKendry, P. (2002b), Energy production from biomass (part 2): conversion technologies, *Bioresource Technology*, 83(1): 47-54.

McKendry, P. (2002c.) Energy production from biomass (part 3): gasification technologies, *Bioresource Technology*, 83(1): 55-63.

Mouillot, F., A. Narasimha, Y. Balkanski, J.F. Lamarque, and C.B. Field (2006), Global carbon emissions from biomass burning in the 20th century, *Geophysical Research Letters*, 33(1).

Oros, D. R., and B. R. T. Simoneit (2001), Identification and emission factors of molecular tracers in organic aerosols from biomass burning Part 1. Temperate climate conifers, *Applied Geochemistry*, 16(13), 1513-1544.

Pan, J. L., and N. Voulvoulis (2007), The role of mechanical and biological treatment in reducing methane emissions from landfill disposal of municipal solid waste in the United Kingdom, *Journal of the Air & Waste Management Association, 57*(2), 155-163.

Pier, P. A., and J. M. Kelly (1997), Measured and estimated methane and carbon dioxide emissions from sawdust waste in the Tennessee Valley under alternative management strategies, *Bioresource Technology, 61*(3), 213-220.

Pronobis, M. (2006), The influence of biomass co-combustion on boiler fouling and efficiency, *Fuel, 85*(4), 474-480.

Reddy, MS, and C Venkataraman (2002), Inventory of aerosol and sulphur dioxide emissions from India. Part II - biomass combustion, *Atmos. Environ., 36*(4): 699-712.

Reiner, T., D. Sprung, C. Jost, et al. (2001), Chemical characterization of pollution layers over the tropical Indian Ocean. Signatures of emissions from biomass and fossil fuel burning, *Journal of Geophysical Research-Atmospheres, 106*(D22), 28497-28510.

Schmid, O., P. Artaxo, W. P. Arnott, et al. (2006), Spectral light absorption by ambient aerosols influenced by biomass burning in the Amazon Basin. I: Comparison and field calibration of absorption measurement techniques, *Atmospheric Chemistry and Physics, 6*, 3443-3462.

Swithenbank, J., V. Nasserzadeh, R. Taib, et al. (1997), Incineration of wastes in novel high-efficiency tumbling and rotating, fluidized bed incinerator, *Journal of Environmental Engineering-Asce, 123*(10), 1047-1052.

Suzuki, T, Y Ikumi, S Okamoto, I Watanabe, N Fujitake, and H Otsuka. (2004), Aerobic composting of chips from clear-cut trees with various co-materials, *Bioresource Technology, 95*(2): 121-128.

Tilman, D., J. Hill, and C. Lehman (2006), Carbon-negative biofuels from low-input high-diversity grassland biomass, *Science, 314*(5805), 1598-1600.

Turn, S. Q., B. M. Jenkins, J. C. Chow, et al. (1997), Elemental characterization of particulate matter emitted from biomass burning: Wind tunnel derived source profiles for herbaceous and wood fuels, *Journal of Geophysical Research-Atmospheres, 102*(D3), 3683-3699.

Wiedinmyer, C., B. Quayle, C. Geron, et al. (2006a), Estimating emissions from fires in North America for air quality modeling, *Atmospheric Environment, 40*(19), 3419-3432.

Wiedinmyer, C., X. X. Tie, A. Guenther, et al. (2006b), Future changes in biogenic isoprene emissions: How might they affect regional and global atmospheric chemistry?, *Earth Interactions, 10*.

Winiwarter, W., and K. Rypdal (2001), Assessing the uncertainty associated, with national greenhouse gas emission inventories: a case study for Austria, *Atmospheric Environment, 35*(32), 5425-5440.

US EPA (1998), Compilation of Air Pollutant Emission Factors, AP-42, 5th Edition, Volume 1: *Stationary Point and Area Sources* Chapter 2: Solid Waste Disposal, Section 2.4, U.S. EPA Supplement E, November 1998. p. 2.4 - 4.

US EPA (2000), Municipal solid waste in the United States: 2000 Facts and Figures. 2000 Update, EPA530-R-02-001.

USEPA (2007), Technology transfer network clearinghouse for inventories & emissions factors, edited.

USEPA (2008), Climate leaders: Greenhouse gas inventory protocol core module guidance - Direct emissions from stationary combustions sources, EPA430-K-08-003, www.epa.gov/climateleaders, May 2008.

11

Industrial Emission Treatment Technologies

Manh Hoang and Anita J. Hill
CSIRO
Australia

1. Introduction

Industrial emissions are becoming one of the most significant environmental issues facing industries. Gaseous emissions from industrial operations can adversely affect the atmosphere (e.g. carbon dioxide CO_2, nitrous oxide N_2O, volatile organic compounds VOCs, steam) and health of people living in surrounding neighbourhoods (e.g. odour, particulates, heavy metals).

The burning of fossil fuels and biomass is the most significant source of air pollutants such as sulphur dioxide SO_2, carbon monoxide CO, nitrogen oxides NO and N_2O (known collectively as NOx), VOCs and some heavy metals. The main source of SO_2 emission is from the combustion of sulphur-containing fuels. Sulphurs are present in coal in the form of pyrites, sulphates or organic sulphurs. In some deposits, the sulphur content can be as high as 4%. Upon combustion, most of the sulphur is converted to SO_2, with a small amount being further oxidized to sulphur trioxide SO_3. In the absence of a catalyst, the formation of SO_3 is slow; over 98% of the combusted sulphur is in the form of SO_2.

The burning of fossil fuels is also the major anthropogenic source of CO_2, one of the important greenhouse gases. Coal is the world's most abundant and widely distributed fossil fuel source. Although many alternatives to combustion of coal are being considered, coal will remain as a principal component of the global energy mix for decades. The International Energy Agency expects a 43% increase in its use from 2000 to 2020 (World Nuclear Association, 2011).

Nitrous oxide N_2O with a Global Warming Potential impact factor of 310 CO_2 equivalent is formed as a by-product from the production of adipic and nitric acid, and nitric acid production is currently believed to be the largest industrial source of N_2O emissions contributing to the ozone layer depletion (US EPA, 2010).

Particulate emission is becoming one of the main causes of respiratory disease worldwide. Sources of particulates can be natural or man-made. Some particulates occur naturally, originating from volcanoes, bush fires, dust storms or by human activities such as the burning of fossil fuels in power plants, vehicles, construction and various industrial processes. Increased levels of fine particles in the air affect lung function and in some cases can cause heart disease.

Steam is the most universal energy carrier. Its application is wide spread and it can be found in all aspects of industrial process. Water vapour is the major driver of greenhouse gas-induced climate change (SMH, 2007). The biggest steam user is thermal power stations where steam is used to generate electricity. The steam consumption in a typical thermal power station of 1000MW capacity is about 2,800 t/h which translates to about 800 kg condensate per second (Hoang, 2011). Waste steam can be found in almost every plant/factory where steam is used; from big industrial establishments such as refineries, power plants, chemical factories, steel makers, ore mining, to medium and small plants such as sugar mills and food processing plants.

Odour is also one of the most significant air quality issues facing industry. As the public and regulators are highly sensitive to nuisance odours, the release of any odorous gas/vapour from industry, whether it may or may not represent a health risk, will result in complaints and possibly regulatory fines. Odour is caused by emission of mixtures of chemicals from a wide range of industrial operations, including pulp and paper, chemical manufacturing, refining, mineral processing, paint, plastics, agrichemical manufacturing, sewage treatment, meat works and rendering plants, food processing, and skin and hide processing at tanneries. Prolonged exposure to foul odours usually generates undesirable reactions in the human body such as discomfort, sensory irritation, headaches, respiratory problems, and vomiting.

Stricter environmental regulation in the industrialized countries has triggered the introduction of cleaner technologies in the last decade. The integration of flue gas desulphurisation, low-NOx, Integrated Gasification Combined Cycle (IGCC), and Pressurised Fluidised Bed Combustion (PFBC) enable coal combustion to occur with higher thermal efficiencies and lower emissions. The combustion of natural gas for electricity generation, transportation and other applications such as industrial boilers has significantly reduced CO_2, SO_2, NOx and particulate emissions. New carbon capture, storage and sequestration technologies are at the stage of economic feasibility evaluation. Research is focused on deploying large scale demonstration projects that provide a pathway for industry to adopt the technologies at full scale. The three primary separation technologies under research, development and deployment are based on liquid absorption, membranes, and solid adsorption.

In the transport sector, the introduction of catalytic converters, reformulated fuels and the improvement of engine performance have also led to a significant reduction in harmful emissions.

Clean technology research for industrial emission treatment is focussed on technologies that offer the highest efficiencies and lowest emissions. These efforts are directed at research, development and deployment of near zero emission fuel conversion, steam utilisation and recycling, odour reduction technologies as well as the development of renewable energy technologies.

2. Prevention and control of industrial emissions

2.1 Emission control technologies and practices

Various technologies exist that have been designed to remove or treat harmful substances from industrial vent gas. These technologies represent a varying degree of commercial readiness. Tens of thousands of hours of operation under industrial conditions have been proven for some technologies while others have only recently been demonstrated at commercial plants.

2.1.1 CO_2 control technologies

Post combustion capture (PCC) is a process that separates and captures CO_2 from a large volume of low pressure flue gases. The majority of demonstration projects rely on liquid absorption using a liquid such as ammonia or another amine, although membrane based processes are in demonstration phase, and research toward solid adsorbent technologies is gaining pace. For each of these separation technologies, research is directed towards high performance, efficient materials and processes to achieve cost effective CO_2 capture for the lowest energy penalty. Carbon capture technologies have the potential to dramatically reduce greenhouse gas emissions from coal-fired power stations if they can be deployed as high speed capture technologies with high selectivity and minimal energy loss.

PCC technologies offer flexibility in that they can be retrofitted to existing power plants presenting the opportunity to utilise the massive historical capital investment in existing plants. PCC technologies can also be integrated with new infrastructure, and renewable technologies can be integrated with PCC. In addition, PCC can be used to capture CO_2 from a range of industrial sources.

Liquid absorbents:

In the PCC process, flue gases from the power station are typically cooled then fed into a CO_2 absorber as shown in Figure 1. The process of capturing CO_2 using aqueous amines has been practiced for over 80 years (Rochelle, 2009) in the removal of CO_2 from natural gas and in the production of beverage-grade CO_2. In the process shown in Figure 1, flue gas that contains CO_2 is contacted with an aqueous amine or ammonia or novel solvent in an absorber column at a relatively low temperature. The CO_2 is absorbed by the solvent. The CO_2 loaded solvent is pumped to a stripping column where it is heated to regenerate the solvent and release the CO_2. The hot regenerated solvent is then returned to the absorber column through a heat exchanger, which cools the hot solvent and preheats the CO_2 loaded solvent going to the stripper. This process has some limitations that include solvent loss, corrosion, and the high energy demand for regeneration of the solvent. Indeed, most of the energy required to capture CO_2 is consumed in heating the CO_2 loaded solvent in the stripper column. It is estimated that a 500 MW_e power plant requires on the order of 1000 metric tons of absorbent in order to separate more than 80 per cent of the CO_2 from flue gas (DOE, 2010).

Research and development activities for solvent based PCC technologies include the synthesis and scale up of novel solvents with higher selectivity and capacity, using waste heat to increase solvent capacity and capture rate, improving the chemical and thermal stability of solvents, and new process designs to reduce footprint. Solvent systems that are responsive to physical and chemical switches other than the traditional thermal and pressure swings are also under investigation. One class of novel solvents under investigation is ionic liquids or low temperature molten salts. They are composed of cations and anions and are liquids at ambient conditions. They are non-volatile, hence avoiding solvent loss, and non-flammable. Their chemical structure can be designed so that they absorb CO_2 and require a low energy for the desorption of CO_2 (Gurkan, 2010; Wang, 2010).

Membranes:

Membrane based separation processes for CO_2, although a less mature technology than the aqueous amine solvent technology, have some advantages over solvent technologies for separating gases. These advantageous aspects include a simple operation with no moving parts, a technology that builds on existing low-cost technology that is already proven at similar scales in other industries - e.g. desalination, an environmentally benign separation technology without the use of hazardous chemicals, a compact and modular technology with a small footprint, and finally a low energy technology (DOE, 2010; Merkel, 2010). Figure 2 illustrates the small footprint of a membrane based gas separation unit for the removal of CO_2 from natural gas as compared to an amine absorption unit. The amine absorption unit was decommissioned because the columns were corroding over time due to the aggressive nature of the amine solution used to adsorb the CO_2. The membrane unit is dwarfed by the amine unit, although both technologies are employed to perform the same separation.

Challenges for membrane based separation include particulate matter and its impact on membrane lifetime, and due to the youth of this technology, there is a lack of operating experience in the power industry. Recent demonstration trials of membrane technology by Membrane Technology and Research Inc (MTR) have been completed in collaboration with the Arizona Public Service Cholla power plant for PCC with the process diagram shown in Figure 3.

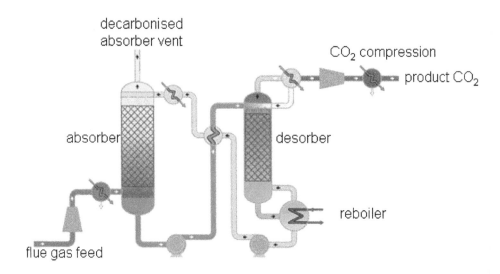

Fig. 1. Solvent based CO_2 capture process flow diagram (adapted from Siemens, 2008).

Fig. 2. Comparison of footprint for solvent versus membrane process units for CO_2 separation (adapted from Fleming, 2009).

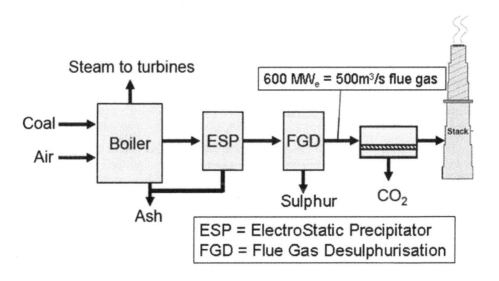

Fig. 3. Membrane based CO_2 capture process flow diagram (adapted from Merkel, 2010).

Performance data for the MTR Polaris™ membrane have been reported for temperatures ranging from 25-50°C (Merkel, 2010). These recent trials have indicated that 95% of the CO_2 from flue gas can be separated at a parasitic energy penalty of 15% using the Polaris™ membrane technology, which can be compared to the estimated solvent technology parasitic energy penalty of 30% (Merkel, 2010). A parasitic energy penalty is indicative of the additional fuel that must be burned (to enable carbon capture) for the same power output. Current material and process efficiency improvements for both technologies aim toward < 10% parasitic energy penalty.

Membranes made from thermally stable polymers offer attractive features such as dimensional stability at elevated temperature, resistance to oxidation and thermal degradation, and stability to harsh chemical environments. A new membrane based on thermally rearranged (TR) polymers has been reported that can perform CO_2 separation processes at ambient temperatures as well as at operation temperatures as high as 230°C (Park, 2010). The gas transport in these TR polymer membranes occurs through a pore architecture that consists of micropores (<2 nm) that allow fast diffusion of gases connected by ultramicropores (<0.7 nm) that are responsible for the molecular sieving. The method of creating this efficient pore architecture in thermally stable glassy polymers is via thermal rearrangement, hence the name TR polymer membranes. The pore architecture can be tuned for particular gas and vapour separations, hence TR polymer membranes are currently in research and development stage for numerous commercial applications including CO_2 capture.

Solid sorbents:

Solid adsorbents utilized in chemical separation applications include molecular sieves, activated carbons, and zeolites, all of which are porous materials with high surface area, typically up to 3000 m² g⁻¹. Pressure swing adsorption (PSA) is a process that relies on preferential adsorption of the CO_2 from gas mixtures when exposed to the porous materials at a high pressure. The pressure is subsequently reduced and the CO_2 is desorbed from the porous solid. PSA technology has relatively low energy requirements and capital costs. Much current research focuses on the development of regenerable solid adsorbents with chemical structures and pore architectures that are designed to increase the selectivity and adsorption capacity for CO_2 and lower the energy consumption in the sorption-desorption cycle. Similar to the research directions in solvent absorption, solid adsorbent materials that are responsive to physical and chemical switches other than thermal or pressure swings, e.g. electrical, magnetic, mechanical, or light stimulation, or the presence of particular chemical species to facilitate either uptake or release of CO_2 are also under investigation.

Metal organic frameworks (MOFs) are one of the topics of current research focus. MOFs are materials with extremely high surface areas, up to 10,000 m²g⁻¹, and their chemistry can be tailored to control the porosity and capacity for CO_2. They are synthesized using metal atoms or clusters linked in a periodic fashion via organic linker molecules to form crystalline porous structures. Figure 4 illustrates the pore architecture of a class of MOF that uses imidazolate organic linkers to achieve a pore architecture that resembles an hourglass with large cages separated by a constriction which results from the asymmetric connectivity of the imidazolate ring (Thornton, 2011). These frameworks are known as zeolitic imidazolate frameworks (ZIFs) and they have been shown through modelling and experiment to offer great promise for CO_2 separations (DOE, 2010). Another characteristic of the best performing candidate CO_2 adsorbent MOFs is that they possess a high density of accessible open metal sites (Yazaydın, 2009). Combining the ability to tailor chemistry in order to tune the affinity for a particular gas, along with the ability to architecture the porosity of membranes and solid adsorbents, is an area of much current research.

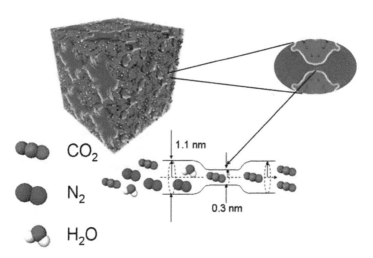

Fig. 4. Schematic pore architecture of a class of MOF known as zeolitic imidazolate frameworks (ZIFs) illustrating the hourglass shaped porosity in which the larger cages provide efficient access for gas molecules to the selective constrictions (adapted from Thornton, 2011).

2.1.2 NOx control technologies for nitric acid plants

Nitrous oxide (N_2O) is a by-product of the manufacture of nitric acid by the Ostwald process. This process involves three basic operations: (i) the combustion of ammonia into nitric oxide, (ii) the oxidation of nitric oxide into nitrogen dioxide and (iii) the absorption of nitrogen dioxide in water to produce nitric acid (Perez-Ramirez, 2003; Maurer, 2005).

Depending the location in the process, the approaches of lowering N_2O from nitric acid plants can be classified into three groups:

- Primary: preventing N_2O being formed in the ammonia burner
- Secondary: removal of N_2O from NOx gases between the ammonia converter and the absorption column.
- Tertiary: removal of N_2O from the tail gas downstream.

The development of N_2O abatement systems aims at the achievement of high efficiency (>90% N_2O conversion) and selectivity (<0.2% NO loss) (Perez-Ramirez, 2003).

Technologies for primary abatement:

The catalytic oxidation of ammonia over platinum-rhodium gauzes takes place through several reaction paths depending on the temperature. Under current industrial operating conditions, the process yields up to 97% NO, the desired product in nitric acid production. Although the process is well established, there are still research efforts in improving the selectivity not only for production yield but also for lowering N_2O emission. Thus optimising the NO selectivity by improving the catalyst, reactor design and operating conditions are considered as part of primary abatement. Examples of efforts toward optimisation include changes in the geometry and configurations of gauzes as well as the development of oxide-based catalysts (Groves, 2006; Perez-Ramirez, 2004a; Perez-Ramirez, 2006a).

Technologies for secondary abatement:

Thermal decomposition:

Norsk Hydro developed a reactor technology based on the thermal gas phase decomposition occurring down-stream from the reactor chamber. The technology has been implemented in an acid plant in Porsgumn Norway in 1990 (Perez-Ramirez, 2004b).

Catalytic decomposition:

Nitrous oxide emissions from nitric acid production are influenced by the degree to which emission control efforts are applied in both new and existing plants (Mainhardt, 1996). The catalytic reduction of N_2O is well known from the open and patent literatures. It was reported that N_2O can be decomposed into N_2 and O_2 over various types of noble metal, metal oxide and mixed oxide catalysts. However there are a limited number of catalysts that have been commercialised. An ideal industrial catalyst must satisfy the N_2O reduction target of >90% under severe oxidising environments at high temperature and pressure and with a minimal loss of NO.

Norsk Hydro patented a series of catalysts comprising $Co_{3-x}M_xO_4$ where M=Fe or Al and x= 0 to 2 supported on CeO_2. The presence of a small amount of ZrO_2 was reported to prevent the degradation of catalyst activity (Nirisen, 2002, 2004). This catalyst is claimed to be non-NO destructive, versatile, active and thermally stable, which can be applied over a wide temperature and gas composition range, e.g. in the presence of oxygen and water vapour. Based on this invention, Norsk Hydro successfully commercialised the catalysts NH-1 and its second generation NH-2.

A Cu-Zn-Al catalyst was developed, patented and commercialised by BASF in 1999 (Schumacher, 1999, 2004). The best catalyst composition was reported as 8% wt CuO, 30% wt ZnO and 62% Al_2O_3. The catalyst was fabricated in star extruded form denoted as O3-80, O3-85 and O3-86.

A Ce-Co catalyst was commercialised by Yara International as Yara 58-Y1. It contains >80% cerium oxide, <1% cobalt (II, III) oxide and <1% aluminium cobalt tetraoxide (MSDS, 2008). This Yara catalyst is claimed to be able to achieve greater than 90% N_2O conversion. This N_2O abatement catalyst technology has been tested and installed in many industrial locations, including 11 Yara plants and 11 locations under the clean development mechanism (CDM) or joint implementation (JI) countries as defined by the Kyoto Protocol (Matthey, 2008).

Johnson Matthey, Yara, Mitsui, Alloy Engineering, Födisch AG, and N.serve have formed a globally alliance called N.serve-Johnson Matthey Alliance in 2008. The N.serve-Johnson Matthey Alliance technology is based on the Yara secondary N_2O abatement catalyst, Yara 58-Y1 and Amoxis® type 10-1R, a lanthanum-cerium perovskite catalyst.

Hermsdorfer Institut Für Technische Keramik (HITK) has patented and produced a ceramic based catalyst consisting of metal oxides such as Cr, Mn, Fe, Co, Ni, Cu and La on a porous ceramic support (Burckhardt, 2000). The patent claims that at temperature above 800°C, 100% N_2O conversion is achievable in laboratory and pilot plant tests.

Apart from the above significant developments, there are a number of catalyst inventions available from the patent literature (Vernooy, 2002a, 2002b; Neveu, 1999; Hamon, 2004; Schwefer, 2001a; Jantsch, 2009; Neumann, 2007; Axon, 2004; Paul, 1978).

Technologies for tertiary abatement:

The tail-gas, leaving the absorber column, contains N_2O, NOx, O_2 and H_2O at temperatures from 523-773K at pressures of 3-13 bar (Perez-Ramirez, 2003).

The composition of the gas at the outlet of the absorber during stable operation varies depending on industrial operation conditions as shown in Table 1 (Perez-Ramirez, 2003).

Gas	Concentration	Typical value
NOx (ppm)	100–3500	200
N_2O (ppm)	300–3500	1500
O_2 (vol.%)	1–4	2.5
H_2O (vol.%)	0.3–2	0.5

Table 1. Composition of the tail-gas at the outlet of the N_2O absorber (balanced by N_2) for a nitric acid plant of 1500 t HNO_3 per day and a tail-gas flow of 200,000Nm^3 h^{-1}.

Major developers/suppliers for tertiary abatement catalysts are Krupp Uhde, ECN/CRI , Yara International, Du Pont, University Jagiellonski, Institute NawozÓw Sztucznych, N.E. Chemcat. Uhde has developed and commercialised a system for the removal of N_2O and NOx from the tail gas of nitric acid plants under the trade name EnviNOx®. The system features a tail gas reactor consisting of an iron zeolite catalyst installed directly upstream of the tail gas turbine. The EnviNOx® catalysts are manufactured by Süd-Chemie with brand names EnviCat®-N_2O and EnviCat®-NOx.

The Uhde EnviNOx® process was first implemented in 2003 in a 1000 Mtpd dual-pressure nitric acid plant by AMI in Linz (Austria). The tail-gas temperature is about 437°C and the applied configuration consisted of a single reactor with two beds of iron-zeolite with intermediate ammonia injection. The presence of nitric oxide in the inlet gas is vital to promote the N_2O decomposition. The reactor in the AMI plant is reported to operate at a stable N_2O conversion of 98% (Hevia, 2008).

The Energy Research Center of the Netherlands (ECN) has developed a combined catalytic technology for the removal of NOx and N_2O based on selective catalytic reduction (SCR) using propane as a reducing agent in 2003. Iron and cobalt-containing zeolites are placed in series in a single reactor operated at pressure up to 10 bar. The conversions of N_2O and NOx ~ 90% were achieved; however, the economic evaluation indicated no cost advantage due to the high dosing of propane (Schwefer, 2001b).

In 2006, ECN tested a trimetallic zeolite-based ECN catalyst prepared by CRI at a pilot scale in a side stream of a nitric acid plant in Ijmuiden, the Netherlands. This N_2O decomposition process has been commercialised in 2007 (Brink, 2002). This catalyst requires no addition of a reducing agent. It can be operated at temperatures of 475-525°C and at pressures of 1-12 bar, with N_2O conversions of 70-95%.

Yara patented a novel ex-framework FeZSM-5 catalyst that led to remarkable performance in simulated tail gases compared to FeZSM-5 catalysts prepared by ion exchange. The multimetallic zeolite and Fe catalyst is used in direct catalytic N_2O decomposition (Perez-Ramirez, 2004b, 2006b).

N.E. Chemcat has developed and commercialised a catalyst named DASH-30D D812. The catalyst consists of palladium supported on alumina silica and magnesium oxide. The catalyst has been installed in numbers of commercial plants by N.E. Chemcat and Sumitomo Engineering including the Caprolactam production plant in Thailand in 2008 (TEI, 2009; CDM 2002), the Pakarab Fertilizer plant in Multan, Pakistan, 2007 (CDM, 2002), the nitric acid plant of the Kaifeng Jinkai Chemical Ind. Co., Ltd, China, 2007 (CDM, 2006a) and the nitric acid plant of Liuzhou Chemical Industry Co., Ltd, China, 2009 (CDM, 2006b). Apart from the above significant developments, there are number of catalyst inventions available from the patent literatures (Russo, 2007).

NOx control technologies for combustion processes:

In combustion processes NOx is formed by the oxidation of fuels that have high nitrogen content such as coal and residual oils. Combustion of low nitrogen content fuels such as distillates or natural gas also produce NOx but in lesser amounts. The most common NOx control technology for the combustion process is low-NOx burners (LNB). They can be used separately or in combination with post-combustion control technologies such as selective non-catalytic reduction (SNCR) or selective catalytic reduction (SCR). This technology involves the modification of the process into a two stage combustion process: a fuel-rich zone where primary combustion takes place and a fuel-lean zone for secondary combustion at lower temperature. LNB technology affords up to 60% reduction for the combustion process (Watts, 2000).

Selective non-catalytic reduction (SNCR) is based on the chemical reduction of NOx molecules into molecular nitrogen (N_2) and water vapour (H_2O) by a nitrogen-based reducing agent such as ammonia or urea. In this process the reducing agent is injected into the post-combustion flue gas stream and heat from the boiler provides the energy for the reduction reaction.

Like SNCR, the selective catalytic reduction (SCR) is based on the reduction of NOx by a reducing reagent except that it is catalysed by a solid catalyst. In this process, a solid catalyst bed is installed downstream of the combustion chamber and ammonia is used as a reducing agent. NOx reductions up to 90% can be achieved by optimising the reactor design and operation such as ammonia dosage and mixing (Hesser, 2005).

2.1.3 SO₂ control technologies

Flue gas desulphurisation:

SO_2 emission is known to have detrimental effects on human health and the environment. SOx emissions are directly linked to the initial sulfur content of the fuel and the combustion parameters do not influence the amount of SOx emitted. To meet emissions regulations, especially when burning high-sulphur coals, it is essential to achieve high levels of SO_2 removal, usually 90% or higher.

The most widely used technology for SO_2 emission control is the flue gas desulphurisation (FGD) process. An extensive review of this technology can be found in the Technical Status Report 012 Cleaner Coal Technology Programme (Kamall, 2000).

FGD technologies can be grouped into wet, dry and regenerable FGD. In wet FGD processes flue gas contacts an alkaline slurry in the absorber. The absorber may take various forms (spray tower or tray tower), depending on the manufacturer and desired process configuration. However, the most often used absorber application is the counterflow vertically oriented spray tower (Kamall, 2000; Srivastava, 2000; Faustine, 2008).

The major reactions occurring in wet FGD are:

i. Absorption

$$SO_2 + H_2O ----> H_2SO_3$$
$$SO_3 + H_2O ----> H_2SO_4$$

ii. Neutralization

$$CaCO_3 + H_2SO_3 ----> CaSO_3 + CO_2 + H_2O$$
$$CaCO_3 + H_2SO_4 ----> CaSO_4 + CO_2 + H_2O$$

iii. Oxidation

$$CaSO_3 + 1/2\ O_2 ----> CaSO_4$$

iv. Crystallization

$$CaSO_4 + 2H_2O ----> CaSO_4 \bullet 2H_2O$$

A relatively high degree of SO_2 removal is usually achieved, with a high level of sorbent utilization. The wet FGD technology has been significantly improved since its first introduction to industry in the 1970s. The new technology features state-of-the-art designs and materials of construction. Highly efficient, compact, and less expensive technology with minimal waste disposal problems has been developed by incorporating oxidation of the calcium sulphite sludge into wallboard-grade gypsum. In addition, wet FGD systems can remove significant particulates due to the contact between gas and liquid phases.

In wet FGD, the limestone forced oxidation (LSFO) is most widely used. In LSFO, air is added to the reaction tank to oxidise the spent slurry to gypsum. The gypsum is removed from the reaction tank prior to the slurry being recycled to the absorber.

In the dry FGD process, SO_2-containing flue gas contacts a lime sorbent. The sorbent can be delivered to flue gas in an aqueous slurry form or as a dry powder. Both methods require dedicated absorber vessels in order for the sorbent to react with SO_2. By-product solids are collected in a dry form along with fly ash from the boiler in the existing particulate removal equipment. Compared with wet FGD systems, SO_2 removal efficiency and sorbent utilization are usually lower. A typical technology of dry FGD is the lime spray dryer (LSD) technology, a dry scrubbing process generally used for low-sulphur coal (Kamall, 2000; Srivastava, 2000; Faustine, 2008).

Hydrodesulphurisation:

The principal operation of a refinery is to convert crude oil into products such as LPG, gasoline, kerosene, diesel, lubricants and feedstocks for petrochemical industries. After separation of the crude oil into different fractions by conventional distillation, these streams are transformed into products with high values by a variety of catalytic processes such as hydrogenation, isomerization, romatization, alkylation, cracking and hydrotreating. Hydrodesulphurisation (HDS) is a catalytic hydrotreating process. It is applied to remove sulphur from natural gas and from refined petroleum products.

The primary objective of refiners before 1990 was to maximise the conversion of heavy oils into gasoline and middle distillates. In the refining processes, the purpose of removing sulphur is to protect the noble metal catalyst from poisoning. The restriction in harmful substance emissions to the atmosphere has spurred the focus on the production of cleaner products. Significant capital investments in the refining industries are necessary to produce cleaner middle distillates to meet environmental standards. Thus another important reason for removing sulphur is to eliminate the SO_2 emission that results from burning refinery products. In a typical industrial hydrodesulphurisation unit, the hydrodesulphurization reaction takes place in a fixed-bed reactor at elevated temperature, 300 to 400°C, and at pressures from 30 to 130atm (CEP, 2009). Industrial catalysts for this purpose consist of cobalt-molybdenum or nickel-molybdenum supported on alumina.

2.1.4 Particulate control technologies

A variety of particulate removal technologies, with different physical and economic characteristics are available.

Electrostatic precipitator (ESP):

This process removes particles by using an electrostatic field to attract the particles onto the electrodes (Rankin, 2011). Once the particles are collected, they are removed through a hopper. ESPs are especially efficient in collecting fine particulates and can also capture trace emissions of some toxic metals with an efficiency of 99% (World Bank Group, 1998). They can operate at elevated temperatures with a very low pressure drop. ESPs have been used in numerous industries such as alumina refineries and cement production plants.

Filters and dust collectors:

Baghouses collect dust by passing flue gases through a fabric that acts as a filter. Depending on the application, various types of filter media including woven fabric, plastic, ceramic, and metallic are commercially available. The flue gas temperature determines the operating temperature of the baghouse gas and the choice of fabric. Fabric filters are efficient (99.9% removal) for both high and low concentrations of particles but are suitable only for dry and free-flowing particles.

Wet scrubbers:

This process relies on the removal of dust particles from a gas stream by liquid phase. A wet-scrubbing technology is used where particulates cannot be removed easily in a dry form; however, the technology generates an effluent that needs to be treated.

2.1.5 Odour control technologies

Odour problems are complex issues, and finding a suitable odour control technique is a challenge as there are numerous options available including physical, chemical and biological treatments. Typical industrial odour emissions are shown in Table 2. High process efficiencies are required for a number of reasons but mainly because humans are very sensitive to low concentrations of odorous substances. The odour thresholds of some substances are shown in Table 3.

Industry	Odorous compounds emitted
Pulp & paper	H_2S, mercaptans, DMS, DMDS
Petroleum	Mercaptans, phenolic compounds, aldehydes
Iron & Steel manufacture	H_2S
Oil & Gas extraction	H_2S, other reduced sulphur compounds
Waste water treatment plants	H_2S, NH_3
Chemical	H_2S, mercaptans, phenols, ammonia, organics
Tannery	Reduced sulphides
Solid Waste Landfill sites	H_2S, decaying organics
Organic fertilisers production	Reduced sulphides, amines

Table 2. Typical industrial odour emissions.

Compound	Odour Detection Threshold (ppm)	Odour Type
Hydrogen sulphide	0.0005	Rotten egg
Mercaptans	0.0001-0.001	Garlic, rotten cabbage
Dimethyl sulphide	0.001	Decayed cabbage
Dimethyl disulphide	0.16-25	Decayed organic matter
Ammonia	55	
Trimethylamine	0.00021	Rotten fish
Methyl amine	0.021	Fishy

Table 3. Detection thresholds for various odorous compounds.

There is a number of available technologies but their applicability is dependant on the characteristics and the physical and chemical properties of the vent gas (Schleglmich, 2005; Busca, 2003).

Wet scrubbing removes odorous compounds by effectively dissolving the vent gas in an aqueous solution. This method can be extremely effective but may transfer the problem from air to waste water.

Adsorption of odorous compounds onto materials with a large surface (eg. activated carbon) offers one of the lowest capital costs. The adsorption media has to be replaced or regenerated based on the amount of odorous compounds removed. As with wet scrubbing, regeneration of adsorbent with steam may transfer the problem from air to waste water. Another problem with adsorption is that the vented gas needs to be free of oil and water as this may clog pores thus reduce the effectiveness of odour removal.

Biofiltration involves passing the odorous gas through a bed of material (e.g. soil, bark, organic mater, synthetic packing). The odorous compounds are adsorbed onto the surface of the bed of material and are then subsequently broken down by microbial action. Biofiltration is suitable for large gas flow rates, and it has low operating costs, but it requires a very large foot print.

Thermal oxidation, at temperatures up to 800°C, of odorous compounds can be very effective but the capital and operating costs are high. Additional air quality problems may arise from the combustion products such as SOx and NOx (Busca, 2003).

Catalytic oxidation of odorous gas offers significant advantages over thermal oxidation, by either speeding up the oxidation reaction or by lowering the required reaction temperature (Busca, 2003; Kahn, 2000). The Commonwealth Scientific & Industrial Research Organisation (CSIRO) has developed versatile catalysts that have been tested and found effective at removing odorous compounds from vapour streams (O'Neill, 2005). These catalysts operate at very low temperature, between 100-300°C, providing significant operational savings and a reduced carbon footprint. The technology is very versatile, it can accommodate very low to high gas flowrates. This technology can be extended to various industrial operations including food processing and preparation, pulp & paper manufacturing, petrochemical industries, chemical manufacturing, cement manufacturing, minerals processing and waste water treatment. The technology is being commercialised by Catalytic Solutions International (CSI) Pty Ltd, based in Perth Western Australia (Figure 5).

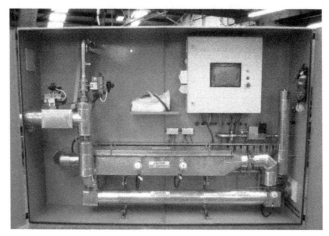

Fig. 5. Catalytic pilot reactor at final completion (Catalytic Solutions International Pty Ltd – Australia, 2011).

2.1.6 Industrial steam emission control technologies

Steam is almost exclusively produced in boilers, the efficiency of which is about 70-80%. Steam is also generated as a by-product of processes such as an evaporator, or when water is used as the cooling medium. Steam generation is an energy intensive process. Fuel cost is the main component in the cost of steam production. Other factors such as the water inlet temperature and the pressure and temperature of the product steam also affect the cost of steam generation (CADDET Energy Efficiency, 2001).

In an industrial process, after transferring its energy, the pressure and temperature of the steam drop significantly. During the process, it is contaminated with volatile chemicals and gases such as air and carbon dioxide. A common practice to deal with spent steam is to use a condenser to collect the water or to discharge the steam to the atmosphere. Discharging the spent steam to atmosphere is not only an energy loss — it is at the same time an environmental issue.

With a higher energy cost and a growing concern regarding environmental effects, it is highly desirable to recover the energy loss by recycling the spent steam. The first step in this process is to remove or separate the steam from other gaseous and volatile impurities.

The energy recovery from waste steam can be partially accomplished by using a heat exchanger for multi-effect evaporation when possible or by a vapour recompression process to make the steam usable for other purposes. For example, in addition to the recommendation of using feedwater economisers for waste heat recovery (Steam Tip Sheet 3, 2006), the US DOE has considered the following approaches to waste steam recovery (Steam Tip Sheet 13, 14, 29, 2006):

- Send low-grade waste steam to power absorption chillers.
- Use steam jet ejectors or thermocompressors to reduce venting of low pressure steam.
- Use vapour recompression to recover low pressure waste steam.
- Use a vent condenser to recover flash steam energy.

Vapour recompression appears to be the most advanced technique. It is comprised of (i) thermal vapour recompression and (ii) mechanical vapour compression, with mechanical vapour compression being the favoured option. Limitations of the current recompression techniques include:

- Additional energy is needed to boost the waste steam to higher potential.
- They cannot handle "dirty" steam, i.e. steam containing impurities. The presence of air, especially at higher volume concentrations, can affect the operation of the vapour recompression process. The reprocessed steam still contains gases and impurities as the process cannot eliminate them.

Industrial interest has stimulated numerous investigations into methods of separating and recovering spent steam for both economical and environmental benefits.

Membranes are commonly used in the steam reforming process, where hydrogen is separated from the exit gas mixture using hydrogen selective membranes. In this process, as well as in the fuel cell context, although the membrane is in contact with the steam, it is not used for steam separation.

RASIRC reported a range of steam processing products based on Nafion™ membrane (Spiegelman, 2009), for example:

- Steam generating units where the function of the membrane is to purify the steam to a level suitable for semiconductor industry or other special applications.
- Gas drying units where water vapour is removed from a gas stream.
- Humidity controlled units where water vapour is introduced into a gas stream using membranes.

A Hewlett-Packard patent (Hewlett-Packard, 1989) "Water vapour permeable material for drying gases" reports a method of drying a gas steam using a membrane comprised of a fluoro-carbon polymer containing lithium sulphonate groups. The polymer is reported to be stable under steam conditions.

CSIRO has patented a membrane and process for steam separation and recovery based on a hydrophilic membrane system (CSIRO, 2011a). The technology has been demonstrated at pilot scale.

2.2 Advanced technology developments

Industrialisation offers benefits but also creates pollution. Industrial and environmental sustainability requires new technologies for clean, renewable energy, renewable feed-stocks, cleaner production, and water and air management. Due to society and business awakening, as well as government incentive schemes, new technologies are being developed that embody engineering, economic and environmental considerations.

2.2.1 Materials for industrial emission control

Energy use and emissions:

Generating power close to where it is needed, or managing when and how energy is delivered and used, can improve energy efficiency, minimise energy consumption, and reduce peak demand on the electricity grid. Energy efficiency improvements proportionally reduce industrial emissions.

There have been significant research and development investments in:

- Materials and processes for lower energy water filtration and desalination.
- New methods to generate oxygen with lower power requirements.
- New methods for producing oxygen at large scales required for power production and leaner combustion processes.
- Greener manufacturing and energy production processes with materials that produce less air and water pollution, and capture pollution before it reaches our atmosphere.
- Efficient fuel combustion in both industry and transport sectors
- Low energy processes for capturing CO_2

The availability of appropriate materials is often critical to the success of new engineering, technology, or processing activities. There will be significant energy saving opportunities through improved materials in reaction unit operations, coupling reactions and separations into hybrid unit operations e.g. syngas separation; hydrogen, carbon monoxide, and carbon dioxide separations, a more economical separation for oxygen enabled combustion, and the use of membrane based separation in place of thermal distillation.

The separation of light olefins from respective saturated hydrocarbons on a commercial scale is performed almost exclusively by cryogenic distillation. This technology is highly energy intensive and there is a strong economic incentive to explore alternative separation technologies.

Catalysis is a critical component of a modern industrialised economy. Catalytic processes are responsible for as much as 90% of chemical production processes (Scott, 1999). Catalysis is currently playing a major role in pollution abatement and prevention, leading to significant industrial waste reduction. During the last decade, significant improvements have been made in energy efficiency for chemical and petrochemical processes, and catalysts have played a major role in these improvements. Cleaner technologies directly address the cause of emissions during the production process by using higher selective catalysts or by the development of new catalysts to enable the replacement of environmentally harmful inputs. Due to the ready availability of n-butane as a natural gas liquid, together with the imposition of strict pollution control measures on atmospheric benzene emission, the replacement of benzene by n-butane as a feedstock for maleic anhydride has been commercialised using a vanadium phosphorous mixed oxide catalyst.

In the membrane development area, most current research and development focuses on new membrane materials; however, research and development efforts are also directed at controlling specific process conditions and process stream compositions in order to take advantage of existing membrane materials, for example the use of membranes to overcome thermodynamic limitations like azeotropes.

Membranes are selective either by pore size or by their affinity for the permeating components. The intrinsic performance characteristics of a membrane material are its selectivity and flux. These characteristics are controlled by the membrane material composition, surface structure and morphology and its compatibility to industrial process operations including the nature and chemical composition of the process stream. Large amounts of research have been devoted to gas separations in recent years such as the separation of hydrogen and hydrocarbons in gas recovery unit operations and high-temperature air separation via ceramic membranes with applications in fuel cells and oxygen-rich combustions. Advanced air separation techniques are constantly being re-engineered with primary aims at raising overall operating efficiencies and increasing the production capacity (EPB, 2009). In numerous industries significant energy savings could be achieved if existing air-fueled combustions could be replaced by oxygen-fueled ones. For this to occur, new energy efficient O_2/N_2 separation technologies need to be developed.

Modern transportation is an essential part of our lives. Forefront materials research has offered significant advantages for this sector by:

- improving fuel combustion efficiency,
- introduction of hybrid vehicles, renewable fuels,
- improving fuel cell-powered vehicle performance,
- hydrogen storage, and
- biofuel production.

One current area of research focus for storage materials for future transportation fuels, such as hydrogen and methane, includes the metal organic frameworks (MOFs) discussed earlier in section 2.1.1. Several MOFs have been commercialized by Sigma-Aldrich and manufactured by BASF (Jacoby, 2008). MOFs suitable for hydrogen storage are being developed for use in fuel-cell-powered automobiles. Optimising the pore architecture, heat of adsorption, the gravimetric capacity, and operational stability of the MOFs are all key criteria for translation of this technology from the laboratory to pilot scale and finally deployment (Sumida, 2009).

2.2.2 Clean and renewable energies

Clean and renewable energies are parts of a long term plan to reshape our economy. Materials have played a major role in:

- Solar power
- Wind power
- Hydrogen production solar power
- Biofuels
- Fuels from lignocelluloses, organic wastes
- Energy from waste steams at higher temperatures and pressures

There has been significant growth in photovoltaics (PV) over the past decade and the cost of electricity from this technology continues to decrease. The increased efficiency has been achieved primarily from advances in material development. However there are still materials research and development challenges for PV technologies including the need to continue to increase solar cell efficiency by improving material properties and cell designs. This increase can be achieved by identifying or developing alternate materials e.g. organic photovoltiacs (CSIRO, 2011b) and developing novel nanoscale surfaces to increase capture of the full spectrum of sunlight.

Advanced materials and process research, development, and deployment are contributing to today's energy technologies as well as helping to meet the challenges of future energy needs.

3. Conclusion

All industries are moving toward more efficient and low or zero emission processes due to economic benefit, regulation, and concern for health and the environment. New materials and process technologies are playing an increasing role in emission treatment and by-product utilisation. This chapter has focussed on technologies that offer the highest efficiencies and lowest emissions including physical and chemical mechanisms for near-zero emission fuel conversion and industrial commodity production such as cement or via refinery operation, steam utilisation and recycling, and odour reduction. Candidate technologies covered in this chapter include selective catalytic reduction techniques and chemical separation technologies such as absorption, membrane and adsorption based processes. Several research topics were identified for focus, ranging from fundamental modelling and theory linked to rapid screening and optimisation to speed material and process development through to process conditions suitable for scale-up and deployment. Low energy, efficient, reliable, and environmentally benign emission treatment technologies are one goal of modern process scientists and engineers, with the future holding the promise of closed loop production.

4. Acknowledgment

It has been the authors' privilege and pleasure to work with the Materials for Energy, Water, and Environment team as well as our national and international collaborators. We thank our colleagues and collaborators for their contribution to this work. We thank Irene Poon of CSIRO Process Science and Engineering for editorial assistance.

5. References

Axon, S., Coupland, D., Foy, J., Ridland, J. & Wishart, I., (2004), *Ammonia Oxidation Process*, WO 2004/096703 A2.

Brink, R., Booneveld, S., Pels, J., Bruijn, F., Gent, M. & Smit, A., (2002), *Combined catalytic removal of NOx and N₂O in a single reactor from the tail gas of a nitric acid plant : final report*. ECN Report ECN-C-02-009, Novem-project no.: 358510/0810, 2002.

Burckhardt, W., Seifert, F. & Winterstein, G., (2000), WO 00/13789, 2000.

Busca, G. & Pistarino C., (2003), Technologies for the abatement of sulphide compounds from gaseous streams: a comparative overview, *Journal of Loss Prevention in the Process Industries* Vol.16, pp.363-371.

CADDET Energy Efficiency, (2001), Maxi Brochure 13 " Saving Energy with Steam Production and Distribution", http://home.iprimus.com.au/nissenr/mb_13.pdf, accessed on 25 August 2011.

Catalytic Solutions International Pty Ltd – Australia, (2011), Odour control technology commercialised, http://intranet.csiro.au/intranet/communication/internalcomm/mondaymail/2010/MM100628/htm/odour.htm, accessed on 25 August 2011.

CDM, (2002), http://cdm.unfccc.int/UserManagement/FileStorage/IB1P5OQ0CDIPDK66E76P7BVS6IZ1YK, accessed on August 20, 2011.

CDM, (2006a), http://cdm.unfccc.int/UserManagement/FileStorage/Y8EP1I75XRZSEDZ3V44KY9IJTIWUXG, accessed on August 17, 2011.

CDM, (2006b), http://cdm.unfccc.int/UserManagement/FileStorage/J52MW58RYOLWAWSHZEWQOI4XV8BSCE, accessed on 17/08/2011.

Chemical Engineering Processing, (2009), http://chemeng-processing.blogspot.com/2009/01/hydrodesulfurization.html, accessed 25 August 2011.

CSIRO, (2011a), Science: Securing the Future, http://www.csiro.au/files/files/py1b.pdf, accessed on 24 August 2011.

CSIRO, (2011b), Low cost energy, using organic photovoltaics, http://www.csiro.au/science/Organic-photovoltaics.html , accessed on 30 August 2011.

DOE, (2010), Basic Research Needs for Carbon Capture: Beyond 2020 Report of the Basic Energy Sciences Workshop for Carbon Capture: Beyond 2020, http://www.sc.doe.gov/bes/reports/files/CCB2020_rpt.pdf, accessed on 25 August 2010.

EPB, (2009), Trend Report No. 10: Industrial Gas, 29th International Exhibition Congress in Chemical Engineering, Environmental Protection and Biotechnology, Frankfurt May 2009.

Faustine, C., (2008), Environmental Review of Petroleum Industry Effluents Analysis, Industrial Ecology, Royal Institute of Technology Stockholm.

Fleming, G. (2009), Gas Separation Membranes MEDAL™ Air Liquide, http://www.tnav.be/pages/documents/BMG_GasSep_Medal-AirLiquide_GFleming.pdf accessed 25 August 2011.

Groves M., Maurer R. , Schwefer M. & Siefert R. (2006), Abatement of N2O and NOx from nitric acid plants with the Uhed EnviNOX ® process – Design, Operating Experience and Current development – NITROGEN 2006 International Conference – 14/03/2006.

Gurkan, E., de la Fuente, J., Mindrup, E. M., Ficke, L. E., Goodrich, B. F., Price, E. A., Schneider, W. F., & Brennecke, J. F., (2010), Equimolar CO_2 absorption by anion-functionalized ionic liquids, *J. Am. Chem. Soc.* Vol.132, pp. 2116–2117.

Hamon, C. & Duclos, D., (2004), *Method For The High-Temperature Catalytic Decomposition Of N_2O Into N_2 And O_2*, WO 2004/052512 A1, 2004.

Hesser, M., Lüders, H., & Henning, R-S., (2005), SCR Technology for NOx Reduction: Series Experience and State of Development, DEER Conference, 21–25/08/2005.

Hevia, M.A.G. & Pérez-Ramírez J., (2008), Assessment of the low-temperature EnviNOx (R) variant for catalytic N2O abatement over steam-activated FeZSM-5, *Applied Catalysis B: Environmental* Vol.77, pp.248-254.

Hewlett-Packard, (1989), *Water-vapour permeable material*, United States Patent, WO89/02447A.

Hoang M. & Nguyen, C., (2011), Membrane and process for steam separation, purification and recovery – US Patent: US 2011/0107911A1, 12/05/2011.

Jacoby, M., (2008), Heading to market with MOFs: for metal-organic frameworks, lab-scale research is brisk as commercialization begins, *Chemical and Engineering News*, Vol.86, pp. 13-16.

Jantsch, U., Lund, J., Gorywoda, M. & Kraus, M., (2009), *Catalyst for the decomposition of N_2O in the ostwald* ,US 2009/0130010 A1, 2009.

Kamall, R., (2000), Technical Status Report 012 on Cleaner Coal Technology Programme. http://webarchive.nationalarchives.gov.uk/+/http://www.berr.gov.uk/files/file 19291.pdf, accessed 20 August 2011.

Khan F.I. & Ghoshal A.K., (2000), Removal of volatile organic compounds from polluted air, *Journal of Loss Prevention in the Process Industries* Vol.13 pp.527-545.

Mainhardt, H. & Kruger, D. (1996) N2O Emission from Adipic Acid and Nitric Acid Production, 1996, IPCC Guidelines for National Greenhouse Gas Inventories, USA.

Matthey, (2008), http://www.noble.matthey.com/pdfs-uploaded/N.serve%20Johnson%20Matthey%20Alliance_Brochure_Sept%202008.pdf, accessed on August 20, 2011.

Maurer R. and Groves M., (2005) Technical research paper No:1, IFA Technical Committee Meeting, 11-13 April 2005, Alexandria, Egypt.

Merkel, T. C., Lin, H., Wei, X. & Baker, R. (2010), Power plant post-combustion carbon dioxide capture: An opportunity for membranes, *Journal of Membrane Science*, Volume 359, pp. 126-139.

MSDS, (2008), MSDS for N2O Abatement Catalyst 58-Y1, 58-Y1-S, by Yara International, 2008.

Neumann, J., Isopova, L., Pinaeva, L., Kulikovskaya, N. & Zolotarskii, I., (2007), Catalyst And Process For The Decomposition Of Nitrous Oxide As Well As Process And Device In Nitric Acid Preparation, WO 2007/104403 A1, 2007.

Neveu, G., (1999), *Method For Reducing Nitrous Oxide In Gases And Corresponding Catalysts*, WO 99/64139, 1999.

Nirisen, Ø., Schoffel, K., Waller, D. & Ovrebo, D., (2002), *Catalyst For Decomposing Nitrous Oxide And Method For Performing Processes Comprising Formation Of Nitrous Oxide*, WO 02/02230 A1, 2002, to Norsk Hydro.

Nirisen, Ø., Schoffel, K., Waller, D. & Ovrebo, D., (2004), *Catalyst for decomposing nitrous oxide and method for performing processes comprising formation of nitrous oxide* , US 2004/0023796 A1, 2004.

O'Neill, G. (2005), Change in the wind, CSIRO Solve Issue 2 Feb 05
   http://www.solve.csiro.au/0205/solve0205.pdf accessed 25 August 2011.
Park, H. B., Han, S. H., Jung, C. H., Lee, Y. M., Hill, A. J., (2010), Thermally rearranged (TR)
   polymer membranes for CO_2 separation, *J Membrane Sci.*, Vol.359, pp.11-24.
Paul, A. S., (1978), *Catalyst for the decomposition of nitrogen oxides*, 4088604, 1978.
Perez-Ramirez J., Kapteijin F., Schoffel K., & Moulijn J.A., (2003), Formation and control of
   N2O in nitric acid production. Where do we stand today?, *Applied Catalysis B:
   Environmental,* Vol 44, pp.117-151.
Perez-Ramirez, J., (2004a), *Method For Preparation And Activation Of Multimetallic Zeolite
   Catalysts, A Catalyst Composition And Application For N_2O Abatement*, WO
   2004/047960 A1, 2004, assigned to Yara.
Pérez-Ramírez, J., (2004b), *Procede Pour La Preparation Et L'activation De Catalyseurs Sur
   Zeolites Multimetalliques, Composition De Catalyseurs Et Application De Ce Procede Pour
   La Reduction Du N_2O*, Pérez-Ramírez, J., (2004b), WO 2004/047960 A1, 2004,
   assigned to Yara.
Pérez-Ramírez, J., (2006a), *Method for preparation and activation of multimetallic zeolite catalysts,
   a catalyst composition and application for N_2O abatement*, US 2006/0088469 A1, 2006,
   assigned to Yara.
Perez-Ramirez, J., (2006b), *Method And System For Guaranteeing The Privacy Of The User
   Identification*, WO 2006/047960 A1, 2006, assigned to Yara.
Rankin, W. J., (2011). *Minerals, Metals and Sustainability*, Ch 12 Management of wastes from
   primary production, Ch 17 Towards zero waste, CSIRO and CRC Press, ISBN:
   9780643097261, Melbourne.
Rochelle, G. T., (2009), Amine scrubbing for CO_2 capture, *Science*, Vol 325, pp. 1652–1654.
Russo N., Fino D., Saracco G., & Specchia V., (2007), N2O catalytic decomposition over
   various spinel-type oxides, *Catalysis Today*, Vol.119 pp. 228-232.
Schleglmich M., Steese J. & Stegmann R., (2005) Odour management and treatment
   technologies: An overview, *Waste Management* Vol.25 pp.928-939
Schumacher, V., Bűrger, G., Fetzer, T., Baier, M. & Hesse, M., (1999), WO 99/55621, 1999.
Schumacher, V., Bűrger, G., Fetzer, T., Baier, M. & Hesse, M., (2004), *Method for the catalytic
   decomposition of N2O*, US 6743404 B1, 2004.
Schwefer, M., Szonn, E. & Turek, T., (2001a), *Method for Removal of NOx and N_2O*, WO
   01/51181, 2001.
Schwefer, M., Maurer, R. & Turek, T. (2001b), *An Apparatus And A Process For Removing N_2O
   During Nitric Acid Production*, WO 01/51415 A1, 2001.
Scott, S., (1999), The role of university research in catalysis innovation, *Canadian Chemical
   News*, July/August p.16.
Siemens, (2008),
   http://www.energy.siemens.com/mx/pool/hq/power-generation/power-
   plants/carbon-capture-solutions/post-combustion-carbon-capture/development-
   of-an-economic-carbon-capture-process.pdf, accessed on 25 August 2011.
SMH, (2007),
   http://www.smh.com.au/news/environment/greenhouse-
   gases/2007/04/02/1175366158796.html, accessed on 17 August 2011.
Spiegelman, J., (2009), Steamer versus pyrolytic torch in photovoltaic manufacturing – a cost
   of ownership comparison,
   http://www.globalsolartechnology.com/documents/solar_2.3_low_res.pdf,
   accessed on 25 August 2011.

Srivastava, R. K., (2000), Controlling SO$_2$ Emissions: A Review of Technologies, U.S. Environmental Protection Agency - EPA/600/R-00/093, November 2000.

Steam Tip Sheet #3, January 2006
    http://www1.eere.energy.gov/industry/bestpractices/pdfs/steam3_recovery.pdf, accessed on 25 August 2011.

Steam Tip Sheet #13, Revised May 2006
    http://www1.eere.energy.gov/industry/bestpractices/pdfs/39315.pdf, accessed on 25 August 2011.

Steam Tip Sheet #14 , January 2006
    http://www1.eere.energy.gov/industry/bestpractices/pdfs/steam14_chillers.pdf, accessed on 25 August 2011.

Steam Tip Sheet #29, January 2006
    http://www1.eere.energy.gov/industry/bestpractices/pdfs/steam29_use_steam.pdf, accessed on 25 August 2011.

Sumida, K., Hill, M. R., Horike, S., Dailly, A., & Long, J. R., (2009) Synthesis and Hydrogen Storage Properties of Be$_{12}$(OH)$_{12}$(1,3,5-benzenetribenzoate)$_4$, *J. Am. Chem. Soc.*, Vol.131, pp. 15120–15121.

TEI, (2009),
    http://www.tei.or.th/Event/eip/080709-ICS-CDM-N2O%20Abatement%20in%20Tail%20Gas2.pdf, accessed on August 20, 2011.

Thornton, A. W., (2011), private communication.

U.S. Environmental Protection Agency report, (2010), "Available and Emerging Technologies for Reducing Greenhouse Gas Emissions from the Nitric Acid Production Industry" December 2010;
    http://www.epa.gov/nsr/ghgdocs/nitricacid.pdf accessed on August 20, 2011.

Vernooy, P.D., (2002a), *Zirconia catalysts for nitrous oxide abatement*, US 2002/0123424 A1, 2002.

Vernooy, P.D. (2002b), *Catalyst for the decomposition of nitrous oxide*, US 6429168 B1, 2002.

Wang, C., Luo, H., Jiang, D., Li, H. & Dai, S., (2010), Carbon Dioxide Capture by Superbase-Derived Protic Ionic Liquids, *Angew. Chem. Int. Ed.* Vol.49, pp. 5978 –5981.

Watts, J.U., Mann, A.N. & Russell, D.L., (2000), An Overview of NOx Control Technologies Demonstrated under the Department of Energy's Clean Coal Technology Program, U.S. Department of Energy, Federal Energy Technology Centre.

World Bank Group, (1998), Airborne Particulate Matter: Pollution Prevention and Control, Pollution Prevention and Abatement Handbook, World Bank Group, July 1998.

World Nuclear Association, (2011), "Clean coal" Technologies, Carbon Capture & Sequestration – World Nuclear Association – April 2011 http://world-nuclear.org/info/inf83.html accessed on August 20, 2011.

Yazaydın, A. O., Snurr, R. Q., Park, T-H., Kyoungmoo Koh, K., Liu, J., LeVan, M. D., Benin, A. I., Jakubczak, P., Lanuza, M., Galloway, D. B., Low, J. J., & Willis, R. R., (2009), Screening of Metal-Organic Frameworks for Carbon Dioxide Capture from Flue Gas Using a Combined Experimental and Modeling Approach, *J. Am. Chem. Soc.*, Vol.131, pp. 18198–18199.

Permissions

The contributors of this book come from diverse backgrounds, making this book a truly international effort. This book will bring forth new frontiers with its revolutionizing research information and detailed analysis of the nascent developments around the world.

We would like to thank Xiao-Ying Yu, for lending her expertise to make the book truly unique. She has played a crucial role in the development of this book. Without her invaluable contribution this book wouldn't have been possible. She has made vital efforts to compile up to date information on the varied aspects of this subject to make this book a valuable addition to the collection of many professionals and students.

This book was conceptualized with the vision of imparting up-to-date information and advanced data in this field. To ensure the same, a matchless editorial board was set up. Every individual on the board went through rigorous rounds of assessment to prove their worth. After which they invested a large part of their time researching and compiling the most relevant data for our readers. Conferences and sessions were held from time to time between the editorial board and the contributing authors to present the data in the most comprehensible form. The editorial team has worked tirelessly to provide valuable and valid information to help people across the globe.

Every chapter published in this book has been scrutinized by our experts. Their significance has been extensively debated. The topics covered herein carry significant findings which will fuel the growth of the discipline. They may even be implemented as practical applications or may be referred to as a beginning point for another development. Chapters in this book were first published by InTech; hereby published with permission under the Creative Commons Attribution License or equivalent.

The editorial board has been involved in producing this book since its inception. They have spent rigorous hours researching and exploring the diverse topics which have resulted in the successful publishing of this book. They have passed on their knowledge of decades through this book. To expedite this challenging task, the publisher supported the team at every step. A small team of assistant editors was also appointed to further simplify the editing procedure and attain best results for the readers.

Our editorial team has been hand-picked from every corner of the world. Their multi-ethnicity adds dynamic inputs to the discussions which result in innovative outcomes. These outcomes are then further discussed with the researchers and contributors who give their valuable feedback and opinion regarding the same. The feedback is then collaborated with the researches and they are edited in a comprehensive manner to aid the understanding of the subject.

Apart from the editorial board, the designing team has also invested a significant amount of their time in understanding the subject and creating the most relevant covers. They scrutinized every image to scout for the most suitable representation of the subject and create an appropriate cover for the book.

The publishing team has been involved in this book since its early stages. They were actively engaged in every process, be it collecting the data, connecting with the contributors or procuring relevant information. The team has been an ardent support to the editorial, designing and production team. Their endless efforts to recruit the best for this project, has resulted in the accomplishment of this book. They are a veteran in the field of academics and their pool of knowledge is as vast as their experience in printing. Their expertise and guidance has proved useful at every step. Their uncompromising quality standards have made this book an exceptional effort. Their encouragement from time to time has been an inspiration for everyone.

The publisher and the editorial board hope that this book will prove to be a valuable piece of knowledge for researchers, students, practitioners and scholars across the globe.

List of Contributors

Giulliana Mondelli, Heraldo Luiz Giacheti and Vagner Roberto Elis
Institute for Technological Research of São Paulo State,São Paulo State University, University of São Paulo, São Paulo, Brazil

Cristina Pomposiello, Cristina Dapeña, Alicia Favetto and Pamela Boujon
Instituto de Geocronología y Geología Isotópica (INGEIS, CONICET-UBA), Argentina

Kamarudin Samuding, Mohd Tadza Abdul Rahman, Lakam Mejus and Roslanzairi Mostapa
Malaysian Nuclear Agency (Nuclear Malaysia), Bangi, Kajang, Selangor, Malaysia

Ismail Abustan
School of Civil Engineering, Universiti Sains Malaysia, Nibong Tebal, Penang, Malaysia

Yifeng Wang, Huizhen Gao, Andy Miller and Phillip Pohl
Sandia National Laboratories, USA

Jan Derco and Michal Melicher
Institute of Chemical and Environmental Engineering, Slovak University of Technology, Bratislava, Slovak Republic

Angelika Kassai
Water Research Institute, Bratislava, Slovak Republic

Daisuke Sugiyama
Nuclear Technology Research Laboratory, Central Research Institute of Electric Power Industry, Japan

Keita Okuyama and Kenji Noshita
Hitachi Research Laboratory, Hitachi, Ltd., Japan

Iheoma Mary Adekunle, Oke Oguns, Philip D. Shekwolo and Augustine O. O. Igbuku
Remediation Department, Shell Petroleum Development Company (Nigeria) Limited, Port Harcourt, Nigeria

Olayinka O. Ogunkoya
University Liaison, Shell Petroleum Development Company (Nigeria) Limited, Port Harcourt, Nigeria

María del Carmen Mingorance Rodríguez
Universidad de La Laguna, Spain

James G. Droppo and Xiao-Ying Yu
Pacific Northwest National Laboratory, USA

Manh Hoang and Anita J. Hill
CSIRO, Australia

Printed in the USA
CPSIA information can be obtained
at www.ICGtesting.com
JSHW011812301024
72690JS00002B/62

9 781632 401304